The Dead Sea Jesus

A Critical Study of the Qumran Scrolls

by

Fernando Klein

Fernando Klein

The Dead Sea Jesus
A Critical Study of the Qumran Scrolls

Copyright © 2010 by Fernando Klein

All rights reserved.

Printed in the United States of America. No part of this book may be used or reproduced in any manner whatsoever without written permission except in the case of brief quotations embodied in critical articles and reviews.

Published by
Fifth Estate, Post Office Box 116,
Blountsville, AL 35031

First Edition
Cover Designed by An Quigley

Printed on acid-free paper
Library of Congress Control No: 2010931482
ISBN: 9781933580982

Fifth Estate, 2010

Fernando Klein

To my loving wife Marcela

**Dedicated to My Mentors,
Too Numerous to Mention,
But Especially
Peter Flint
James Vandekam
and
Martin G. Abegg**

Fernando Klein

Author's Preface

The object of this work is primarily to furnish students with a handy text-book, which, it is hoped, will facilitate the study of the particular texts. But it is also hoped that the book will be acceptable to the general reader who may be interested in the subjects with which they deal.

It has been thought advisable, as a general rule, to restrict the notes and comments; indeed, it is much to be desired that these translations may have the effect of inducing readers to study the larger works.

My principal aim, in a word, is to make some difficult texts, important for the study of the Dead Sea Scrolls, more generally accessible in faithful and scholarly translations. In most cases these texts are not available in a cheap and handy form.

Fernando Klein is Professor of Middle East Religions and Chair of the Religious Studies Department at Complutense University, Madrid, Spain. He has published several books on the Bible, Judaism and Christendom, including "The Naked Bible", "The Gnostic Gospels", "The Church of Paper", and so on.

Fernando Klein

Contents

Prologue

1. Placement of Qumran

2. Finding and Backgrounds

3. Origin of the Qumran Community

4. Identification of the Community

5. The Doctrine of the Qumran Community

    a. Introduction
    b. Theological Beliefs
    c. Prohibition against Use of Oil
    d. No Private Property
    e. The Pure Meal
    f. Prohibition against Spitting
    g. Strictness of Sabbath Observance
    h. Oaths upon Entrance
    i. Secretiveness
    j. Entrance Procedure
    k. The Entrance to the Sect
    l. The Qumran Calendar
    m. The War
    n. Qumran's two Messiahs

6. Archeology of Qumran

    a. Dating the findings
    b. Physic´ Space of the Community: Historical Evolution
    c. The Manuscripts of the Complex

7. Judaism in Qumran

    a. Temple and Priesthood
    b. Synagogue and Prayer
    c. The Centrality of Torah
    d. Apocalyptic Eschatology

8. Was Christ in Qumran?

    a. Introduction

b. Historical Jesus and the Community of the Yahad
c. Conclusions

9. The Stone that Killed God

a. "Hazon Gabriel" or "The Vision of Gabriel"
b. Polemic Contents
c. Transcription of the text in English
d. Strong Criticism
e. Some Other "Shaken Stones": the fragment 7Q5 and 7Q4

The Scrolls From the Dead Sea

Survey of the Caves

Rewritten Bible Texts

Genesis Apocryphon (Tales of the Patriarchs (1QapGen=1Q20)
A Genesis Florilegium (4Q252)
Reworked Pentateuch (4QPPa=4Q158)
Enoch and the Watchers (4Q227)
Enoch (Hanokh, 4Q201)
The Book of Giants (4Q203, 1Q23, 2Q26, 4Q530-532, 6Q8)
Exhortation based on the Flood (4QFloodAp=4Q370)
Exodus 4Q22 (paleo Exodm), Column 1
Leviticus (Va-Yikrah, 11Q1 PaleoLev)
Joshua Apocryphon (4Q522)
A Biblical Chronology (4Q559)
Hur and Miriam (4Q544)
Pseudo-jubilees (4Q227)
Aramaic Tobit (4Q196)
Stories from the Persian Court (4Q550)

Legal and Ritual texts

Phylacteries
Ritual Purity Laws (4QTohorota=4Q274)
A Baptismal Liturgy (4Q414)
Acts of Torah (4Q394)
Various Laws (4Q159)
Songs of Sabbath Sacrifice (Shirot `Olat ha-Shabbat, 4Q403 ShirShabbd)
Some Torah Precepts (Miqsat Ma`ase ha-Torah, 4Q396 MMTc)
Tehillim (11Qp1)
The First Letter on Works Reckoned as Righteousness (4Q394-398)
The Second Letter on Works Reckoned as Righteousness (4Q397 - 399)

A Pleasing Fragrance (Halakhah A-4Q251)
Mourning, Seminal Emissions, etc. (Purity Laws Type A-4Q274)
Laws of the Red Heifer (Purity Laws Type B-4Q276-277)
The Foundations of Righteousness (The End of the Damascus Document: An Excommunication Text-4Q266)

Commentaries

Isaiah Commentary (4QIsaiah Pesher, 4Q161, 4QpIsa)
Habakkuk Commentary (Pesher Habakkuk, 1QP HAB)
Hosea Commentary (Pesher Hoshe`a, 4Q166 4QpHosa)
Zephaniah Commentary (4QZephaniah Pesher 4Q170 a 1:12-13)
Micah Commentary (4QMicah Pesher, 4Q168)
Nahum Commentary (4QNahum Pesher 4Q169)
Targum of Job (4QtgJob, 11QtgJob)

Wisdom Literature

Wisdom Text (1Q26, 4QWisda)
Collection of Proverbs (4QWisd)
Wiles of the Wicked Woman (4Q184)
The Parable of the Bountiful Tree (4Q302a)

Messianic and Visionary Works

The Messiah of Heaven and Earth (4Q521)
The Messianic Leader (Nasi-4Q285)
The Servants of Darkness (4Q471)
The Birth of Noah (4Q 534-536)
Words of Michael (4Qmich, 6Qunidar)
The New Jerusalem (1QJNar, 2QExc, 4QJMa, 5QJNar, 11QJN)
The Tree of Evil (A Fragmentary Apocalypse-4Q458)
Vision of Jacob (4QAJa=4Q537)
Midrash on Last Days (4Q174, 4QFlorilegium)

Prophecy and Apocalyptic

The Chosen One (4Qelect, 4QarNC)
The Book of Secrets (1Q27, 4Q299-301)
The Divine Throne Chariot
The Coming of Melchizedek (11Q13)
Messianic Apocalypse (4Q521)

Prophets and Pseudo-Prophets

The Angels of Mastemoth and the Rule of Belial (4Q390)

Pseudo-Jeremiah (4Q385)
Second Ezekiel (4Q385-389)
Pseudo-Daniel (4Q243-245)
The Son of God (4Q246)
Vision of the Four Kingdoms (4Q547)

Testaments and Admonitions

Aramaic Testament of Levi (4Q2 13 -214)
A Firm Foundation (Aaron A-4Q541)
Testament of Kohath (4Q 542)
Testament of Amram (4Q543, 545-548)
Testament of Naphtali (4Q2 15)
Admonitions to the Sons of Dawn (4Q298)
The Sons of Righteousness (Proverbs-4Q424)
The Demons of Death (Beatitudes-4Q525)

Sectarian literature

War Rule (Serekh ha-Milhamah, 11Q14)
The Community Rule (Serkeh ha-Yahad)
Damascus Document (Brit Damesek)
Tongues of Fire (1Q29, 4Q376)
The Copper Scroll (3QTreasurea)
Prayer for King Jonathan (Tefillah li-Shlomo shel Yonatan ha-Melekh, 4Q448)

Hymns, Psalms and Poetry

The Chariots of Glory (4Q2 8 6-2 87)
Hymns of the Poor (4Q434, 436)
Thanksgiving Hymns 1QH)
Apocryphal Psalms (4QPsf=4Q88, 4QapPs=4Q448, 11QPsa-b=11Q5-6)
The Thanksgiving Psalms
Plea for Deliverance (11QPsa=11Q5 col.xix)
The Children of Salvation (Yesha') and The Mystery of Existence (4Q416, 418)

Calendrical Texts And Priestly Courses

Priestly Courses I (4Q3 2 1)
Priestly Courses II(4Q32 0)
Priestly Courses III-Aemilius Kills (4Q323-324A-B)
Priestly Courses IV (4Q325)
Heavenly Concordances (Otot-4Q319A)
Calendrical Document (Mishmarot)

Divination, Magic and Miscellaneous

Brontologion (4Q318)
Physiognomic Horoscopes (4QCryptic-4Q186, 4QPhysiogn=4Q561)
An Amulet Formula Against Evil Spirits (4Q 560)
The Era of Light is Coming (4Q462)
He Loved His Bodily Emissions (A Record of Sectarian Discipline-4Q477)

Afterthought

1. Qumran and the historical sources

2. Qumran and the final rebellion against the Roman Empire

    a. Introduction
    b. The Final War
    c. Rabbi Akiva
    d. The Letters of Shimo Bar Kokhba

Bibliography

Prologue

The Dead Sea Scrolls comprise a vast collection of Jewish documents written in Hebrew, Aramaic, and Greek, and encompassing many subjects and literary styles. They include manuscripts or fragments of every book in the Hebrew Bible except the Book of Esther, all of them created nearly one thousand years earlier than any previously known biblical manuscripts. The scrolls also contain the earliest existing biblical commentary, on the Book of Habakkuk, and many other writings, among them religious works pertaining to Jewish sects of the time

In this book will be analyzed the last and important discovering in the Dead Sea, specially, "The Vision of Gabriel", and all the scrolls related to the Christendom. "The Vision of Gabriel" or "Hazon Gabriel" were written, using ink, in stone, and introduce us to a critical epoch of Jewish history: it speaks about a person rebelled in front Rome nearly the end of the I century B.C. that was resuscitated at the third day. The point is that some scholars think that this kind of text and premonition were applied to Jesus by their followers or the scribes. As a matter of fact, it is linked to Messiah and that he should die and rise from the dead at the third day, the core of the Christianity faith. This text belongs is a part of a bigger collection of scrolls named the "Dead Sea scrolls, discovered and analyzed at Qumran site, Israel, since 1947.

The legends of what was contained in the Dead Sea Scrolls are far beyond what was actually there. There were no lost books of the Bible or other literature that there was not already other copies of. The vast majority of the Dead Sea Scrolls were simply copies of books of the Old Testament from 250-150 B.C. A copy or portion of nearly every Old Testament book was found in Qumran. There were extra-biblical and apocryphal books found as well, but again, the vast majority of the scrolls were copies of the Hebrew Old Testament. The Dead Sea Scrolls were such an amazing discovery in that the scrolls were in excellent condition and had remained hidden for so long (over 2000 years). The Dead Sea Scrolls can also give us confidence in the reliability of the Old Testament manuscripts since there were minimal differences between the manuscripts that had previously been discovered and those that were found in Qumran.

Clearly the Dead Sea Scrolls and the schism that caused the Dead Sea sect to arise can only be understood in the context of Jewish history and the sectarianism of the Second Temple period in Judea. Across twenty centuries, the Scrolls speak to us of the pluralism that existed in ancient Judaism, each group competing to be the "True Israel" and each claiming a monopoly on the true interpretation of the Torah. The Scrolls provide us with a window into an ancient Jewish time and give us a glimpse of an ancient Jewish sect who resided in the desert on the shores of the Dead Sea. Reflected in the Scrolls are their religious traditions and beliefs, their legal tenets and social structure of the sect.

The most prevalent opinion given by scholars has identified the Qumran sect with the Essenes, of whom Josephus and Philo wrote. While it may be legitimate to attempt to prove Essene authorship as many scholars have done, it is however, illegitimate to use this theory as a universally accepted position on which all Qumran texts are interpreted. Scholarly ethics and integrity, and scientific investigation demand that each text from the caves, along with the Greek writings concerning the Essenes by Philo and Josephus, be subjected to their own separate critical review before conclusions are made.

It must be remembered that Josephus, the primary source of information about the Essenes, wrote primarily for Greek and Roman audiences, and that he wrote approximately two hundred years after the founding of the sect. At this late date, it would be impossible for him to have first hand knowledge. Also, he himself "admits to having included more than one group of sectarians under the heading 'Essenes'.

The Dead Sea Scrolls are a "puzzle": What time frame do they cover, and whom do they refer to? Very few scholars have examined the period from 37 BC to AD 71 as the possible setting for the scrolls. Nevertheless, the scrolls allude to events that only have real relevance in this time period. Even Jesus is not mentioned anywhere in the Scrolls. The Scrolls prove that the Jews were reading and studying texts with a doctrine that could influence in the ideology of early Christendom.

1. Placement of Qumran

The Dead Sea is located in Israel and Jordan, about 15 miles east of Jerusalem. It is extremely deep (averaging about 1,000 feet), salty (some parts containing the highest amount of salts possible), and the lowest body of water in the world. The Dead Sea is supplied by a number of smaller streams, springs, and the Jordan River. Because of its low elevation and its position in a deep basin, the climate of the Dead Sea area is unusual. Its very high evaporation does produce a haze yet its atmospheric humidity is low. Adjacent areas to it are very arid and favorable for the preservation of materials like the Dead Sea Scrolls.

The Bible's description, in Genesis 19, of a destructive earthquake near the Dead Sea area during the time of Abraham is borne out by archaeological and historic investigation. While no evidence remains of the five cities of the plain (Zeboim, Admah, Bela or Zoar, Sodom, and Gomorrah) their sites are believed to be beneath the waters at the southern end of the sea. Archaeological sites near the Dead Sea include Masada, Ein Gedi, and Qumran (where the Dead Sea Scrolls were found).

2. Finding and Backgrounds

In 1947, young Bedouin shepherds, searching for a stray goat in the Judean Desert, entered a long-untouched cave and found jars filled with ancient scrolls. That initial discovery by the Bedouins yielded seven scrolls and began a search that lasted nearly a decade and eventually produced thousands of scroll fragments from eleven caves. During those same years, archaeologists searching for a habitation close to the caves that might help identify the people who deposited the scrolls, excavated the Qumran ruin, a complex of structures located on a barren terrace between the cliffs where the caves are found and the Dead Sea. Within a fairly short time after their discovery, historical, paleographic, and linguistic evidence, as well as carbon-14 dating, established that the scrolls and the Qumran ruin dated from the third century B.C.E. to 68 C.E. They were indeed ancient! Coming from the late Second Temple Period, a time when Jesus of Nazareth lived, they are older than any other surviving biblical manuscripts by almost one thousand years.

Another of the seven scrolls was of a more curious nature. Now named by researchers the "Community Rule" (it was first translated and published under the title "Manual of Discipline"), this large and fairly compete manuscript represented a type of Jewish religious writing previously unknown. It appeared to be a document related to the conduct and beliefs held within a sectarian Jewish community sometime between 150 BCE. and 70 CE. – a community seemingly very much like the Essenes described in antiquity by the Jewish historian Josephus.

In 1949 a team lead by Roland de Vaux (an academic and Dominican priest who would dominate Dead Sea Scroll studies for the next two decades) surveyed the cave at Qumran where the scrolls had been found, discovering pottery shards and several more manuscript fragments. Two years later de Vaux directed archeological excavation of the Khirbet Qumran ruins located just below the cave. Between 1952 and 1956 ten additional caves containing scroll fragments were discovered near Qumran, almost all located by Bedouins who made a business of scouring through the area. The most impressive cache – discovered again by Bedouins working on at Qumran after de Vaux's 1952 expedition – was located in a man-made cave less than 200 yards from Khirbet Qumran. Named "Cave 4" (in order of its discovery), it contained about 15,000 scroll fragments, identified eventually as the remains of 574 separate manuscripts.

In his Ecclesiastical History, Eusebius relates the story of Origen, who consulted scrolls found in caves near Jericho for his "Hexapla," a comprehensive redaction of the Hebrew Scriptures completed in the first half of the third century C.E. *In the . . . edition of the Psalms . . . [Origen reported] again how he found one of [the Texts] at Jericho in a tunnel in the time of Antoninus the son of Severus.*

3. Origin of the Qumran Community

In 199 B.C. Antiochus III (the Great) wrested Palestine from Ptolemy at the Battle of Panion. This marked the beginning of a new era of Jewish history. While the Ptolemies had been tolerant of the Jewish way of life, the Seleucids were determined to enforce Hellenism on the Jews.

The crisis came under Antiochus IV (Epiphanes). He erected a pagan diety on the temple altar. He proclaimed the death penalty for circumcision. Forbid Sabboth observance or the celebration of the Feasts of the Jewish calendar. Copies of the Scriptures were ordered destroyed. An aged scribe name Eleazar was flogged to death for refusing to eat swine.
In the village of Modin, 15 miles west of Jerusalem, Mattathias revolted against Antiochus and killed his emissary. He destroyed the heathen altar and escaped to the hills with his 5 sons. Others joined him in a guerrilla war on the Syrians. He changed the practice of not fighting on the Sabboth if in self- defense.

Shortly after the revolt started, Mattathias died. His third son Judas replaced him. He was known as "the Maccabee" (the hammer). He quickly rallied many Jews to the cause. Soon they marched on Jerusalem and winning victory, entered the temple and removed all the signs of paganism installed there.

Peace was short-lived. The Syrian general, Lysias, defeated the Maccabee in a battle near Jerusalem, and besieged the city. Lysias then maneuvered a peace agreement. Judas left the city and Alcimus became the new Syrian ruler. Alcimus then murdered many of the loyal Jews. Civil war renewed and Judas died in a battle.

Jonathan, a brother of Judas then came to power and through diplomacy became high priest and governor of Judea. His brother, Simon, became governor of the Philistine coastal area. When Jonathan died, Simon, his brother, replaced him.

Simon was old and was the last son of Mattathias, and he passed the kingdom on to John Hyrcanus. Thus started the Hasmonean dynasty. John Hyrcanus expanded the borders of the state on every side. The Maccabean struggle was long over and new rivalries developed. John died in 104 B.C. and the area thrived until Pompey arrived in 63 B.C.

Pompey Besieged Jerusalem for 3 months, then breached the fortifications and killed 12,000 Jews. He allowed temple worship to continue, but Jerusalem was made tributary to the Romans. The last vestige of Jewish independence was removed. Judea was incorporated into the Roman province of Syria and it lost the coastal cities, the district of Samaria, and the non-jewish cities east of the Jordan. Rome ruled with an iron hand and in 37 B.C. installed Herod as king subject to Rome.

4. Identification of the Community

The rise of the Jewish sects is traceable to the impact of Hellenism on the life and culture of the Near East. When the new clashes with the old, violent reactions frequently result. This is particularly true when the new ideology has religious and moral overtones. Many of the Jews were willing to attempt a synthesis of Greek civilization and Hebrew religion. Jews in Palestine as well as Jews throughout the Hellenistic world, adopted Greek names, subscribed to Greek philosophies, and looked to Greek institutions as the harbringers of cultural progress. The Jews of Palestine were generally more conservative than their Greek-speaking cousins in Alexandria and the other great Hellenistic centers, but they were not unaffected. We may assume that these Jews felt that their loyalty to the faith of their fathers was in no way impaired by making peace with the new attitudes which Alexander and his successors had advocated.

Other Jews reacted violently against the Hellenizers. They saw Hellenism as a way of life which was opposed to that prescribed in their Torah. The immodesty of the Greek gymnasium and the neglect of Jewish religious rites by the Hellenistically minded younger generation seemed to indicate trouble. As idolatry had been the besetting sin of Israel before the exile, so Hellenism was regarded as the new temptation to unfaithfulness.

The Jews who reacted against Hellenism are known as the Hasidim (Chasidim) or Assidians. They were, by definition, the party of "the pious." As the Sadducees of the New Testament times continued the basic ideology of the earlier Hellenizers, so the Pharisees and the Essenes sought to preserve the basic tenets of the Hasidim. The Law of God was basic in Hasidic thought. They were willing to suffer martyrdom rather than transgress its precepts. They supported the sons of Mattathias in the early days of the Maccabean revolt, but they left the Hasmoneans as soon as their religious liberties had been won from the Seleucids. Freedom to obey the law was to them an adequate goal, and political independence was quite unnecessary.

It is, however, during the time of the second temple up through the tenth century that is perhaps the most fertile period for the development of Jewish sects. The Roman occupation and the second Diaspora created an environment of great hardship, out of which people tried to escape their pain through religious zeal. As may be supposed, this provided a fertile ground for the growth of religious sects.

The foremost historian of the second temple period was Flavious Josephus. Flavious Josephus is a very controversial figure, and there is much that has been written about the historical accuracy of his accounts. That, however, is another topic. Other primary sources that detail information about sects of the second temple are: the Talmud, Mishna and Tosepta - the 'Abot de Rabbi Nathan, the Christian scriptures. Modern archaeology is also a good source of information on second temple sects.

From Josephus, we know of five major sects: Pharisees, Sadducees, Essenes, Zealots and Sicarii. Josephus divides those sects into three groups: Philosophical (religious), nationalist, and criminal. Of those listed, according to Josephus, the first three are religious:
For there are three philosophical sects among the Jews. The followers of the first of which are the Pharisees; of the second, the Sadducees; and the third sect, which pretends to a severer discipline, are called Essenes (Bellum Judaeum [Wars of the Jews] 1 chapter 8.2).

The Zealots and Sicarii, described as political or criminal factions by Josephus, were groups of extreme nationalists. Their religious beliefs were inseparable from their political convictions. This is not to say that the Pharisees and Sadducees were not political. All of the sects of the second temple were political, and religious; there was no separation between the two.

The Pharisees, Sadducees and Essenes are well known to many people. Most historians, though there is some debate about this, consider the Pharisees to be the forefathers of the Rabbis. The Pharisees were unique in their belief in an oral law, complementary to the Law of Moses in the Torah. This was one of the key differences between the Pharisees and other sects, as Josephus points out in AJ:

"What I would now explain is this, that the Pharisees have delivered to the people a great many observances by succession from their fathers, which are not written in the laws of Moses; and for that reason it is that the Sadducees reject them, and say that we are to esteem those observances to be obligatory which are in the written word, but are not to observe what are derived from the tradition of our forefathers (AJ 13.297)"

The Sadducees are described by Josephus mostly in contrast to the Pharisees. As observed in the passage above, the Sadducees are those who reject the observances not written in the Law of Moses. See also the passages below:

"Now for the Pharisees, they say that some actions, but not all, are the work of fate, and some of them are in our own power, and that they are liable to fate, but are not caused by fate... And for the Sadducees, they take away fate, and say there is no such thing, and that the events of human affairs are not at its disposal; but they suppose that all our actions are in our own power, so that we are ourselves the causes of what is good, and receive what is evil from our own folly (AJ 13, book 5.9)"

In as much as the Sadducees were defined mainly by their strict adherence to the law of Moses as the only legitimate source of Jewish law, they have often been thought of as the progenitors of the Karaites. No conclusive historical evidence, however, exists to prove this contention.

The Essenes are described at length by Josephus in BJ:

"These last [the Essenes] are Jews by birth, and seem to have a greater affection for one another than the other sects have. These Essenes reject pleasures as an evil, but esteem continence, and the conquest over our passions, to be virtue. They neglect wedlock, but choose out other persons children, while they are pliable, and fit for learning, and esteem them to be of their kindred, and form them according to their own manners. They do not absolutely deny the fitness of marriage, and the succession of mankind thereby continued; but they guard against the lascivious behavior of women, and are persuaded that none of them preserve their fidelity to one man (BJ book II 8.2)."

Early in this period of discovery an hypothesis about the source and authors of the scrolls had formed in the minds of de Vaux and his associates. In retrospect, it was only a working hypothesis. But it became a story fixed in history. Faced with several pieces of a puzzle - ancient Hebrew scrolls stored in a cave, a manuscript among those scrolls tentatively identified as the rule of an Essene community, and the ruins of an ancient community's dwelling directly below the cave - de Vaux fit the puzzle's pieces into a temptingly obvious picture: The Dead Sea Scrolls were the library of an Essene community that once occupied the ruins at Khirbet Qumran. Details disclosed from early excavations at Khirbet Qumran all worked neatly into the story: the ruins contained a large room that would have been a scriptorium (a term previously used to describe rooms in medieval monasteries); remnants of long tables were found that could have served for copying lengthy scrolls; and three ink wells were found.

The "Qumran Hypothesis" - attributing the origins and authorship of the scrolls to an Essene community at Khirbet Qumran, a theory perhaps more accurately called the "Qumran-Essene dogma" - became a party line in Dead Sea Scrolls studies for the next 40 years. The integrity of this thesis was buttressed by highly restricted access to the scrolls. Manuscripts were parceled out for study and Text to a small clique of academics, directed by de Vaux.

Fernando Klein

In 1955, literary critic Edmund Wilson published an influential series of articles in The New Yorker magazine (later release in book form) which help cement in popular imagination this accepted story of the Dead Sea Scrolls and their creators, the Essenes who dwelt at Khirbet Qumran. Indeed, Wilson took the tale a tantalizing step further, fleshing out the possibility (broached in 1950 by the French academic André Dupont-Sommer) that the first Christians may have borrowed ideas from the people of the Scrolls. Similar to the first Christians, Wilson explained, the Essenes at Qumran had honored an anointed Teacher of Righteousness, performed ritual washings or "baptisms", and shared a sacred meal. Popular interest in the Scrolls has continued ever since to be stimulated by conjectured links between the Qumran scrolls and early Christianity.

The Essenes were, as I said, a Jewish group known to have been active from the mid-second century BCE to the time of the first Jewish revolt against Rome. We know of their existence mainly from the classical writers Philo, Pliny and Josephus. Scholarly interest in the group was reawakened after the discovery of the Dead Sea Scrolls, when it was thought that the Scrolls may give us further insight into the Essene community. I have been looking at the writings of Josephus (_Jewish War_), Pliny (_Natural History_) and Philo (_Quod Omnis Probus Liber Sit_) and trying to find similarities between their descriptions of the Essene sect and descriptions of the Qumran community in the Dead Sea Scrolls. As well as looking at evidence of the beliefs and practices of the Essene and Qumran Communities, I have considered the archaeological and geographical information in order to try and discover whether the communities were related.

In the last two decades of the twentieth century, several objections to the Qumran-Essene thesis of the Scrolls' origins were voiced within the academic community. Even louder objections arose over continued refusal of the Dead Sea Scrolls "team" to allow all qualified scholars open access to unpublished materials in the collection. After forty years, Scrolls research remained the exclusive domain of a small, self-selected team of scholars. Worse still, over several decades the group had made woefully little progress publishing material from the collection, particularly the large cache of scroll fragments discovered in Cave 4. The whole project was becoming an academic scandal, intermittently punctuated by conspiracy theories suggesting occult purposes motivating sequestration of the yet unpublished materials.

The doctrinal tenants of the Qumran community indicate that they could not have been a group of Sadducees. The Qumran sect had a belief in angels and in the resurrection of the dead, which "was not a central tenet, though they did believe that those who lay in the dust would rise (see 1QH 6:34-35). The '?Children of Light?' are promised eternal life along with the '? Children of Heaven?' (1QS 11:5-9)." The New Testament tells that the Sadducees do not believe in the resurrection, therefore, the sect at Qumran could not have been part of the Sadducees.

Phillip Sigal, former Rabbi of Congregation Ahavas Israel in Grand Rapids and author, says that the Qumranites cannot be identified with the known groups of ancient Israel, but they were "part of a vast complex of groups or movements, all of whom can be subsumed under the rubric 'Pharisees.'" Sigal includes the Essenes, Philo's Therapeutae, and other pietistic, separatist movements as being part of the Pharisees.iii[22] Sigal further states that the Qumran group was an offshoot of the Hasidim who had joined the Maccabean revolt c. 167 B.C., but later withdrew support of the Hasmoneans due to the replacement of the legitimate priesthood by the Hasmonean line and by Simon the Hasmonean's assumption of the title "Prince of the People," which appeared to the Qumranites as usurping David's throne.

Some apparent contradictions show that the Qumranites were not Essenes. The first contradiction to the Essene theory is that the Essenes were reportedly celibate. Josephus, Pliny, and Philo all agree that this was the case, but Josephus also describes a group of Essenes who married. Celibacy was not part of the Hebrew tradition, but there were sectarian groups in Israel who did not marry and refrained from sexual activity. Perhaps those who married did so to procreate, only having sexual relations for that purpose. "These groups were outside the mainstream of Jewish society and lived in a remote region near the Dead Sea. There is no evidence to suggest that celibacy was a requirement for membership in the Qumran community of the Essenes. Three of the unearthed manuscripts, (classified CD, QM and IQSA), believed to be from the community's library, refer to the presence of women and children in the community. The skeletal remains of women were found in the community's cemetery."

5. The Doctrine of the Qumran Community

a. Introduction

There were four main documents found along with the Biblical texts in the caves near Qumran that shed some light on the community life. These documents are the Manual of Discipline, the Damascus Document, the Thanksgiving Hymn, and the Order of Warfare. These documents tell of the regulations and beliefs of the community and from these works "one may deduce that the people ate, lived, and worked together on a common basis." Furthermore, they withdrew from "official" Judaism, retreating to the desert to live a monastic lifestyle of "moral and spiritual discipline." From a sociological view, says that their goal was "to create a utopian society for its members."

The Essenes stressed ethical correctness: moderation and sobriety characterized their whole behavior. We can mention the following practices as being associated with the Essenes:

- Wealth was not accumulated and property was shared.
- Slaves were not allowed.

- Truth was regarded highly, oaths were unnecessary.
- Before each meal a cold bath was taken as part of the purification ceremony.
- White clothes were worn as symbols of purity and sobriety.
- On admission to the group, a member was given three external tokens: a small chopper, an apron and a white robe. According to Philo and Pliny, marriage was rejected and celibacy advocated. Josephus refers, however, to a group of Essenes who allowed marriage (Bell. Jud. II:160f).
- No animal sacrifices were offered, because they had sacrificed themselves for the Lord's cause.
- The communal meal was a high point in the Essene ritual.

We can list the following basic characteristics of the Qumran community:

- Scriptural interpretation was made by the pesher method, in which the historical situation of the passage was applied to their own time and situation.
- Both fulfillment of the law and faith in the Teacher of
- Righteousness are necessary to gain salvation.
- The community was strictly separated from the rest of society who were considered to be the wicked forces of darkness.
- The community was very disciplined and followed many rules and regulations that were intended to protect the purity of the
- community.
- Communal worship consisted of a fellowship meal administered by the priests.
- A strong eschatological orientation pointing toward the dawning of God's kingly rule and a coming war between the forces of light and darkness permeated the community.
- This war would involve both human and supernatural (angelic) forces.
- The community expected the coming of the Prophet and both a priestly and a lay Messiah.
- The communal society was structured by a ranking of priests first, then Levites, laymen, and proselytes.
- The ranking determined where each person would sit during public meetings, and the order in which opinions and questions would be discussed.
- Holiness was an obligation and infractions were punished by various levels of disciplinary action, including being put out of the community for severe offenses.

Different segments of Second Temple Judaism had many religious practices, including certain rituals and dietary practices. The Qumranites preferred to be more restrictive in the areas of "dietary traditions, purity customs, and domestic relations" according to Phillip Sigal. The holy days were observed according to a solar, not lunar calendar (as was used in Jerusalem!), which created disagreements between

Qumran and Jerusalem about when festivals and holy days would be observed. Most notably was the celebration of the day of Pentecost.

Geza Vermes explains that the non-sectarian Jewish calendar was regulated by the lunar movements with monthly variations of twenty-nine or thirty days. This was a 354-day year that needed to be adjusted by adding an additional month every thirty-six lunar months. The Qumran community, seeing that this was an artificial calendar that did not correspond with the four seasons. They rejected the traditional Jewish calendar and used a 364-day calendar based on the sun. This calendar was divisible by the number seven to equal fifty-two weeks per year exactly. The four seasons were 13-weeks long, divided by three months of thirty days each "plus an additional 'remembrance' day." As a matter of fact, many Jews of the second temple period, including the Qumran sect, were fatalistic, believing that their actions were predetermined (at least in part) by God, they were influenced by astrology.

We can compare the prophecy of Jeremiah 31:31-33 with Qumranite theology.

"Behold, days are coming," declares the LORD, "when I will make a new covenant with the house of Israel...But this is the covenant which I will make with the house of Israel after those days," declares the LORD, "?I will put My law within them and on their heart I will write it; and I will be their God, and they shall be My people."

The same ideology of Jeremiah's prophetic utterance was the very core of the Qumran community's basic beliefs. The Essenes at Qumran saw themselves as the final 'remnant' of Israel and the heirs of the 'new Covenant' as was prophesied.

The most distinctive belief held by the Essenes, however, was that they themselves were the community of the new covenant living in the last days immediately prior to the eschatological realization of Israel's hope.

The sect was an extremist offshot of the Jewish apocalyptic movement, whose basic doctrine was the expectation of the soon end of days. When it comes, the wicked would be destroyed, and Israel freed from the yoke of the nations. Before this, God would raise for Himself a community of elect who were destined to be saved from the divine visitation, and who were the nucleus of the society of the future. The Dead Sea Sect carried these views to extremes specific to itself. They believed that God had decreed not only the end but also the division of mankind into two antagonistic camps called "the sons of light and the sons of darkness", lead by superhuman "prince of light" and "angel of darkness" respectively. Reference is also made to "the spirit of truth" and "the spirit of perverseness" which are given to mankind. Of these, each person receives his portion, in accordance with which he is either righteous or wicked. Between these two categories God has set "eternal enmity" which would cease only in the end of days, with the destruction of the spirit of perversion and the purification of the righteous from its influence. Then "the sons of the spirit of truth" would receive their reward.

The bulk of mankind was immersed in evil and liable to suffer divine visitation. To avoid this destiny, members of the sect chose to go to the wilderness and to conduct there a strict way of life in a zealous preparation for future reward. The members of the sect regarded themselves as "an eternal planting", and lived in readiness for the advent of the end of days, when God would raise up for Himself the future Human society, in which they would be "leaders and princes".

The members of the sect may have had several forms of organization. Two of them are described in documents known as the Manual of Discipline and the Damascus Document. The Manual of Discipline called for a full communal life: "they shall eat communally, and bless communally, and take counsel communally". The document does not deal with an event of anyone being born, and the community was presumably a celibate male one. The community strictly observed the laws of ritual purity, regarded all non-members as ritually unclean, and insisted on "obedience from he lower to the higher". For this purpose there was a list of members according to their gradings that was drawn anew every year. The members of the sect were "volunteers" who joined its ranks of their own free will. Offenses against internal discipline were punished, in accordance with the special code, by temporary exclusion.

Due to the Damascus document, however, another form of organization also existed, allowing private property, women and children, and organization as a whole was looser. The sect followed its own interpretation of traditional Judaism which had at least one clear peculiarity: it stated that a calendar of 364 days had to be adopted. They believed to be the only people to know the exact order of the planets and therefore the right calendar. The Temple service was regarded as obligatory, but the sect dissociated itself from the contemporary Temple which its priests, according to them, has defied. In the future the sect would conduct the Temple service in a fitting matter; until that time it exhibited a certain tendency to regard its organization and life as having a religious significance equal to that of the sacrificial service.

The history of the sect and the development of its ideas are unknown. However, some details about its founder are known from documents. He is known as "the teacher of Righteousness", and figures both as the spiritual leader who guides his followers in the path of truth and as the social leader who contends with the ruler in Judea, "the Wicked Priest" (the apparent reference to one of the Hasmonean kings). The Damascus Document states that before the advent of the Teacher of Righteousness there existed a group whose members were "for 20 years like blind men groping their way at noon", until god raised for them the Teaches "to guide in the way of His heart".

b. Theological Beliefs

Josephus distinguishes the three Jewish "philosophies" on the basis of their views on divine sovereignty ("fate") and human freedom; the three formed a spectrum of possibilities: the Pharisees believe that in some things human beings have free will while in others they do not; the Essenes deny human freedom and attribute everything to God's sovereignty ("fate"); the Sadducees reject "fate," holding that human beings have free will.

> At this time there were three sects among the Jews, who had different opinions concerning human actions; the one was called the sect of the Pharisees, another the sect of the Sadducees, and the other the sect of the Essenes. Now for the Pharisees, they say that some actions, but not all, are the work of fate, and some of them are in our own power, and that they are liable to fate, but are not caused by fate. But the sect of the Essenes affirm, that fate governs all things, and that nothing befalls men but what is according to its determination. And for the Sadducees, they take away fate, and say there is no such thing, and that the events of human affairs are not at its disposal; but they suppose that all our actions are in our own power, so that we are ourselves the causes of what is good, and receive what is evil from our own folly. (Ant. 13.171-73)

The depiction of the Essenes as believing "that fate governs all things, and that nothing befalls men but what is according to its determination" agrees with similar assertions in the Qumran sectarian writings. (Unlike Josephus' statements, these assertions are stated, of course, in non-Hellenistic statements.) There are passages in the Qumran sectarian writings that assert God's complete sovereignty over human affairs, including presumably including the choices of human beings (1QH-a 10.9-10; 1QS 1.7-8, 19-20; 11.11, 17-18). Likewise, in several passages from the Qumran sectarian writings, it is affirmed that the eternal destiny of each Jew (or perhaps all human beings) is determined by God. In 1QS 3.18-25, Jews are differentiated according to whether he walks in the spirit of truth or of falsehood. Those "born of Truth" originate from a fountain of light, while those "born of falsehood" originate from a source of darkness. It adds that the sons of righteousness are ruled by the Prince of Light and the sons of falsehood are ruled by the Angel of Darkness. The implication is that Jews find themselves in one of these categories and then live out their lives accordingly. Similarly, in 1QH-a 15.13-19, the author confesses that the inclination of every spirit is in the control of God. Not only did God create the righteous from the womb, in order to bless with salvation, but God also created the wicked from the womb, in order to punish in the day of wrath. Since of the three Jewish "philosophies" only the Essenes hold to the full sovereignty of God ("fate") and since 1QS and 1QH-a, two obviously sectarian writings, affirm that all affairs, human or otherwise, are under the control of God, it seems to follow that the Qumran community was Essene.

Josephus also notes as a distinctive belief of the Essenes their belief in the immortality of the soul and that righteous souls receive a reward of eternal, incorporeal life.

> Indeed, it is a firm belief among them that although bodies are corruptible, and their matter unstable, souls are immortal and endure forever; that, come from subtlest ether, they are entwined with the bodies that serve them as prisons, drawn down as they are by some physical spell; but that when they are freed from the bonds of the flesh, liberated, so to speak, from long slavery, then they rejoice and rise up to the heavenly world (War 2.154-55)

Similarly, according to Ant. 18.18, the Essenes "declare that souls are immortal, and consider it necessary to struggle to obtain the reward of righteousness." (The phrase "reward of righteousness" is a genitive of origin: reward resulting from righteousness.) The reward no doubt is eternal life. Essene convictions about the soul differ from those of the Sadducees insofar as they did not believe in an immortal soul at all and the Pharisees insofar as they believe that the righteous souls are destined "to revive and live again," referring to a renewed corporeal existence (Ant. 18.14; see War 3.374). What Josephus reports about the Essenes' post-mortem hope is consistent with what the fact in the Qumran sectarian writings nothing is said about a renewed corporeal existence for the righteous who have died. It should be noted, however, that Hippolytus claims that the Essenes had a doctrine of the resurrection. He writes, "The doctrine of the resurrection has also derived support from among them, for they acknowledge both that the flesh will rise again and that it will be immortal in the manner as the soul is imperishable." It seems that Hippolytus has interpolated this statement into Josephus' account or the common source to which both he and Josephus were indebted.

c. Prohibition against Use of Oil

Because oil can transmit ritual uncleanness, the Essenes refrained from using it on their bodies, so as not to render themselves ritually impure. This is consistent with the view expressed in the Qumran sectarians writings that oil is easily susceptible to ritual impurity, unlike the more liberal views of their opponents.

| War 2.123 They think that oil is a defilement; and if any one of them be anointed without his own approbation, it is wiped off his body; for they think to be sweaty is a good thing, as they do also to be clothed in white garments. | In 4QMMT it is said that oil poured from one container to another communicates the uncleanness of the oil in the original container and is not pure for being separated. In CD 12.15-17, oil that is found on wood stones or dust is said to |

	communicate uncleanness.

d. No Private Property

In the classical sources, the Essenes are said to have no private property. This agrees with the procedure in the Rule of the Community whereby the property of full members will be "assimilated" into the common resources of the community.

War 2.8.2; 122 "These men are despisers of riches, and so very communicative as raises our admiration. Nor is there any one to be found among them who hath more than another; for it is a law among them, that those who come to them must let what they have be common to the whole order, — insomuch that among them all there is no appearance of poverty, or excess of riches, but every one's possessions are intermingled with every other's possessions; and so there is, as it were, one patrimony among all the brethren." Natural History 5.17.4 (73) "A people ... without money"	1QS 6.18-23 "When he has completed one year within the community, the many shall be asked about his affairs (lit. "things") with regard to his insight and his works in the Torah. If the lot should go out to him, that he should approach the assembly of the community according to the priests and the multitude of the men of their covenant, then both his property and his possessions shall be given to the hand of the man who is the examiner over the possessions of the many. And he shall register it into the account with his hand, and he must not bring it forth for the many. He must not touch the drink of the many until he has completed a second year among the men of the community. When that second year has been completed he shall be examined according to the many. If the lot goes out to him to approach the community, he shall be registered in the order of his rank among his brothers, for Torah, judgment and purity and his property shall be assimilated. His counsel and his judgment shall belong to the community."

e. The Pure Meal

In the classical sources, the Essenes ritually purify themselves by bathing in cold water before they eat a common meal. In the Rule of the Community, regulations for a common meal eaten in a state of ritual purity effected through washing with water are found.

War 2.8.5; 129-31	
"After which they assemble themselves together again into one place; and when they have clothed themselves in white veils, they then bathe their bodies in cold water. And after this purification is over, they every one meet together in an apartment of their own, into which it is not permitted to any of another sect to enter; while they go, after a pure manner, into the dining-room, as into a certain holy temple, and quietly set themselves down; upon which the baker lays them loaves in order; the cook also brings a single plate of one sort of food, and sets it before every one of them; but a priest says grace before the meal; and it is unlawful for any one to taste of the food before grace be said. The same priest, when he has dined, says grace again after the meal; and when they begin, and when they end, they praise God, as he who bestows their food upon them."	1QS 5.13-14 "They shall not enter the water to partake of the pure meal of the holy ones, for they shall not be cleansed unless they turn from their wickedness; for all who transgress his word are unclean." In 1QS 6.13-23, it is stipulated that the candidate for admission into the community is not allowed to touch the food of the pure meal until a year has passed; he must wait another year before he is allowed to drink the wine (also 1QS 6.3-6).

f. Prohibition against Spitting

War 2.8.9; 147	
"Accordingly, if ten of them be sitting together, no one of them will speak while the other nine are against it. They also avoid spitting in the midst of them, or on the right side."	1QS 7.13 "And a man who spits in the midst of the session of the many shall be punished for thirty days."

g. Strictness of Sabbath Observance

War 2.8.9; 147 According to Josephus the Essenes were stricter with respect to the observance of the Sabbath than other Jewish groups. He writes, "Moreover, they are stricter than any other of the Jews in resting from their labors on the seventh day; for they not only get their food ready the day before, that they may not be obliged to kindle a fire on that day, but they will not remove any vessel out of its place, nor go to stool thereon."	Based on the regulations for the Sabbath outlined in CD 10-11, it is clear that the Qumran community was stricter than early rabbinic halakah, which is probably more or less reflective of the Pharisaic position.

h. Oaths upon Entrance

Josephus says that the Essenes take an oath before eating. Similarly, the Rule of the Community requires oaths on those who enter the community.

War 2.139 "But before touching the common food, he makes solemn oaths before his brothers."	1QS 5.7-8 "Everyone who enters into the council of the community shall enter into the covenant of God in the sight of all those who devote themselves. He shall take upon himself a binding oath to return to the Law of Moses, according to all that he has commanded."

i. Secretiveness

Josephus indicates that the Essenes did not reveal to outsiders their distinctive beliefs and practices. The same injunction to secretiveness occurs in the Qumran sectarian writings.

War 2.141 "Also he swears to conceal nothing from the members of the sect and to reveal nothing to outsiders, even though unto death is used against him."	1QS 4.5-6 "Concealing the truth of the mysteries of knowledge" 1QS 9.16-19 "But one must not quarrel with the men of the pit in order that the counsel of the Torah may be

	concealed in the midst of the men of deceit" CD 15.10-11 "Let no one makes the precepts known to him before he stands before the examiner."

j. Entrance Procedure

The procedure by which a prospective member formally becomes a part of the Essenes as described by Josephus (War 2.137-42) is very similar to instruction for the initiation of new members in 1QS 6.14-23.

k. The Entrance to the Sect

The hymns represent the most original literary creation in the Dead Sea Scriptures. It is true that they are, in the main, mosaics of Biblical quotations and that they often exhibit all the learned and tortured exploitation of Scripture that we find later in the medieval poetasters (payye-tanim) of the synagogue. This, however, is merely the trammel of literary convention, and it should no more dull our ears to the underlying passion and authenticity of feeling.

In cases where their beginnings can still be recognized, the hymns often open with the words, 'I give thanks unto Thee, O Lord'. For this reason they have come to be known as the Psalms (or Hymns) of Thanksgiving-a title which is all the more appropriate when it is remembered that in the early synagogue and church, 'thanksgiving' was a technical term for a clearly defined type of liturgical composition. Some of the pieces, however, begin with the alternative formula, 'Blessed art Thou', and this is equally significant in view of the fact that 'blessing' seems likewise to have denoted in antiquity not only a formula of benediction but also a specific type of hymn. Accordingly, the joint title, Blessings and Thanksgivings (Hebrew, Bera-choth we-Hodayoth) would appear to be the most adequate.

The Community Rule is an almost intact document that describes a closely-knit society that above all else seeks purity and correct observance of Jewish Law. It rejects all outsiders, except those seeking admission to the sect. The heart of this document are the legalistic sections. Here the nature and structure of the group are defined, procedures are given for the admission of new members to the sect, and penalties and punishments are outlined for various offenses.

The social character of the group is clearly a very close-knit society in which each man has his function and his rank. "The group lives in a perpetual state of ritual purity, joining together for some meals, as well as maintaining regular sessions for study of scripture and liturgical praise of God. Each man sees himself as a part of this group which in itself constitutes a sanctuary in exile, a replacement Temple, which to the sectarians, was currently in the hands of the evil doers."

From the Community Rule we learn that the Zadokite priests held a superior role in the sect, although a sectarian assembly called the Moshav-ha-Rabbin, made the major decisions. From various portions of this document, the role of the priests seems to be evolving and is in the process of becoming ceremonial. Also located in the Community Rule are passages which speak of withdrawing to the desert to fulfill the command of Isaiah 40: 3 - to prepare a way in the wilderness for the Lord at the end of days.

"When these form a community in Israel, according to these rules they shall be separated from the midst of the settlement of the people of iniquity to go to the desert to clear there the road of the Lord, as it is written, 'In the desert clear the road of the Lord, straighten in the wilderness a highway for our God.' (Isaiah 40: 3) This in the interpretation of the Torah (which) He commanded through Moses to observe, according to everything that is revealed from time to time, and as the prophets have revealed by His holy Spirit. (Rule of the Community 8: 12-16)"

This passage tells us that to prepare a way in the wilderness, specifically means to interpret the Torah. The initiation process speaks of the group's attempt to achieve purity, since they have no Temple. And the penalties and punishments are designed to weed out impurities and contamination from the outside, as they prepare a way in the wilderness for the Lord, who He Himself (according to the War Scroll) will fight the final battle against their opponents, the Sons of Darkness, at the End of Days.

l. The Qumran Calendar

Using the moon to determine Qumran calendar dates in the Dead Sea Scrolls sheds light on the longstanding enigma of when the priestly temple courses served. The Dead Sea Scrolls have provided a wealth of information about religious practices during the first century or so BC in the community at Qumran near the Dead Sea. A surprisingly large portion of the scrolls deals with keeping time, which was essential for knowing exactly when the sacred feasts prescribed in the law of Moses should be celebrated. The scrolls make it clear that the group at Qumran felt that the other Jewish sects were mistaken in the calendar they were using, which was based on the phases of the moon. This was probably one of several reasons that caused them to withdraw from Jerusalem and to celebrate their own feasts at the times they felt were proper according to what has become known as the Qumran calendar.

Some of the scrolls provide long lists of dates and ample detail about the calendar, but apparently no one has correlated the Qumran calendar to our Gregorian calendar. That is, when the scrolls speak of a certain date on which a festival was held, just exactly what date did that correspond to on our calendar? The best correlation to our calendar so far is that the two principal calendar scrolls date to about 50-25 BC, based on the style of handwriting.[2] Fortunately, that is much more precise than the dates ascribed to the entire set of scrolls and the community in general, being from the late second century BC through the destruction of the temple in AD 70.

While several documents discovered at Qumran give schedules of events according to their calendar, the best descriptions of the workings of the calendar itself are probably found in the Book of Jubilees and the Book of Enoch. Although those books are not included in our Bible today, both were held in high regard at Qumran, equal to others we now include in the Old Testament.

The calendar had 364 days each year, beginning on a Wednesday every spring. It had four quarters of exactly 13 weeks each, so that every quarter-year began on a Wednesday. Each quarter had three months, the first two having 30 days, and the third having 31 days. The months were numbered from 1 to 12, beginning in the spring. Thus, it had a feature desired by many modern businessmen: it was so tightly tied to the week that every day occurred on the same day of the week every year. In particular, their sacred feast days always occurred on fixed days.

Of course, having only 364 days, the calendar year was 1.24 days short of a true solar year of 365.24 days. They must have had some method for inserting an extra week often enough to keep this calendar aligned with the seasons, because some of the offerings, such as First Fruits, had to occur at certain seasons of the year, when the barley or wheat would be ripe. While scholars have acknowledged that an intercalation system was needed (meaning inserting extra days to align with the solar year), they have noted that no method is mentioned in the scrolls.

The scrolls also give us one big clue about how the Qumran calendar might have been intercalated. The two most famous and detailed calendar scrolls are called 4Q Calendar Document A and B (also called 4Q320 and 4Q321) because they were found in the 4th cave discovered at Qumran. The latter ("Scroll B") is a continuous 7-year listing of events: one year of dates of lunar observations followed by a 6-year list the dates of both lunar events and also the feast days and how the beginnings of each month aligned with the week and priest cycle. It is clear that every year in that period had exactly 364 days.

The fact that it is a 7-year listing with the last 6 years treated as a separate entity is an important clue. It means they had almost certainly noticed that six years completes a cycle where the same priest would again be officiating on New Year's Day. That is, 6 years of 52 weeks each exactly equals 13 Priest Cycles of 24 weeks (6 x 52 = 13 x 24 = 312 weeks), so the two cycles would begin to repeat.

The current state of understanding is that the Qumran calendar has a year of 364 days comprising 12 months of 30 or 31 days each and has holy days occurring every year on known fixed days. Moreover, the Priest Cycle was uninterrupted for at least one 7-year period. What has not been known is just when that 7-year period occurred, except that it is estimated to have begun about 50-25 BC.

m. The War

Most of the materials of the War Scroll (1QM) originate in older sources, such as the Scriptures and 1 Maccabees, manuals of military tactics, and other texts from Qumran. Behind the War Scroll there might lie a set of beliefs and writings on the theme of the eschatological battle. What interests us here is the nature of the Qumranic redaction. What novel elements did it introduce to the theme of the eschatological battle? The first novelty is the crucial role played by the priests in this war. By blowing their trumpets they direct the actual fighting, even though they must stand apart from the troops in order not to be polluted by blood (1QM IX, 7b-8). The pre-eminent role of the priests is not surprising, the community being of priestly origin and subject to priestly leadership. According to the rules of Yahweh's war, the priests had several duties to perform in the holy war: they were required to offer sacrifices before the people set out to war, and to ask for a divine oracle in order to ascertain whether Yahweh was willing to wage war on the side of his people and grant them victory. They also needed to encourage the soldiers.

On the one hand, the priestly contribution to the war was strengthened in 1QM; on the other hand, the priests who were obliged to follow the commands of the Torah, including strict purity laws (such as 1QM VII,3-7 and IX,7b-8), could not participate in the actual fighting. The crucial role of religion or priests was not typical of Greco-Roman military manuals. In the final analysis, the question arises, whether the tactics described were in fact intended for an actual war at all, or whether the War Scroll was rather a utopian tactical treatise for an eschatological holy war. The treatise implied an idealistic view of how the war should be waged according to the Torah (years of release 1 QM II,8) and how the prescriptions of Torah should be obeyed in wartime (purity laws). The final form of the War Scroll was redacted at Qumran, and it mirrored the views of the community, which lacked the wherewithal to wage war. Therefore, the Sons of Light include other people closely related to the community (I,2b-3), and furthermore, the Sons of Light were to be supported by a host of angels (XII,1-5). The final redaction changed the earlier ethnic war of the sources into something more like a dualistic final battle between the Sons of Light and the Sons of Darkness.

Dualism is the most remarkable novelty represented in the War Scroll in comparison with its sources and the traditions of Yahweh's war in the Scriptures. The dualistic structure is redactional, it is "imposed at a later stage on an earlier non-dualistic war-rule." In a war there are, of course, always two sides fighting against each other. But wars are usually local or national, they do not include the whole world, as in 1QM. In wars of Yahweh it is normal for Israel and its God to form the one side, with another nation or tribe and its god(s) on the other side. Israel and its God belonged together, as did other countries and their god(s). It was assumed that a religion coincided with national or ethnic boundaries. The new dualism implied that in the eschatological battle the battlefront did not run between nations but across the nations; at least the Sons of Darkness did not consist solely of representatives of the other nations, but the "violators of the covenant" joined them (1QM I,2a). The dualism had religious and ethical components, as early Judaism generally emphasized that only a remnant would be saved in future wars and disasters (see for instance Is 10:22-23 and 4QpIsaa II,1-8).

The dualism of the War Scroll was, however, of Persian origin and was cosmic and universalistic in its nature. It offered the potential for loosing religion from its national and ethnic bonds and making it universal in scope, but this potential was not exploited in the War Scroll. What happened instead was a narrowing of religious outlook. The Qumran community separated the Sons of Light from the main body of the people of Israel, and assigned the title of the Sons of Light to the members of their own community and others with similar or related beliefs and halakha (1QM I, 2b-3), while excluding their countrymen who had broken the covenant and were therefore counted among the Sons of Darkness (1QM I,2b). Dualism was adopted at Qumran in order to strengthen the identity of the members of the community. Dualism was an effective weapon for this purpose. Light, goodness and life were confined to the community and its few allies in Israel, while the violators of the covenant, all outsiders and all other nations were counted among the Sons of Darkness. It was like a small group facing the entire world. The story of David and Goliath comes to mind (1QM XI,1-2), but God, and Michael and the host of angels secured victory for the Sons of Light, even though their enemies were such great empires as Assur and Egypt, and the Romans ("kittim").

The Sons of Light were determined to win the battle (1QM I,5.10). According to the final redaction, the battle has seven "lots": in the first three the Sons of Light are stronger, while in the following three lots the Sons of Darkness thrust them back. Finally, in the seventh lot God's mighty hand brings down Belial and his hosts (1QM I,13-15). The Sons of Darkness and everything evil had to be totally destroyed (1QM I,5 and XVIII,1-6). After this the Sons of Light would be able to live in peace and happiness, being blessed by God (1QM I,8-9). This dualistic ideology, which would have had the potential for a universalistic development of religion at Qumran, finally led to a narrowing of outlook. Their sights were not focused on the cosmic, universalistic fight between good and evil, light and darkness, but on the dichotomy between the Sons of Light and the Sons of Darkness among their co-religionists and the various parties and groups. By this sharp dichotomy, the Qumran community endeavoured to strengthen its "sectarian" identity and invoke confidence in its good eschatological lot among the Sons of Light. In the Maccabean era, early Judaism was divided into different religious groups or parties. Dualism was not the reason for these divisions but an effective means of justifying the position taken by a group and detaching their ideology from that of others. This was one part of the process where nationalistic and ethnic identity was narrowed down to a "sect". This particularistic development is already attested in CD and 1QS. Another aspect of dualism was to introduce it into the microcosm of the individual's heart and the inner struggle between the spirit of truth and the wicked spirit. This is the famous doctrine of the two spirits as expounded in the Community Rule (1QS III-IV).

The self-understanding of the community had changed quite substantially since the time when 4QMMT was written. While the Qumran Community of the second century B.C.E. was open to negotiations with other religious groups concerning the right halakha, and in particular with the priestly circles around the Temple (4QMMT C, 1-17 and 25-30), in the final redaction of the War Scroll we encounter a sectarian community making a sharp distinction between the community and outsiders. This development would perhaps be more understandable if the community had experienced a threat to its existence and felt that it lived in a hostile environment without hope of being understood by outsiders or political and religious leaders. They had had bitter experiences from their countrymen's breaking the covenant and joining their adversaries, the Sons of Darkness. The labels such as "The Wicked Priest", "The Man of Lie" and "the traitors of the new covenant" or "violators of the covenant" in 1QpHab (for instance II, 1-6 and VIII, 8-13) demonstrate this atmosphere. Dreadful incidents such as the fate of the Teacher of Righteousness (1QpHab XI,4-8) and the fates of eight hundred Pharisees who were crucified by Alexander Jannaeus (103-76 B.C.E.) were likely to create horror and insecurity in all religious groups, including the Qumran Community (4QpNah I,6-9).

The dualism adopted in 1QM was intolerant in nature but still non-violent in practice, except for the eschatological battle. For this dualism no compromise was allowed. With the representatives of darkness and evil, the sons of Belial, the Prince of Darkness, one must not negotiate; they must be destroyed. It did not even suffice to subjugate the Sons of Darkness to divine rule, although one liturgical section (1QM XII,10b-18, repeated in XIX,1-8) has preserved the notion, often attested in the Scriptures, that other nations must bow down and serve victorious Israel and Judah. It was impossible for a small group, like the Qumran community, to destroy its enemies, the Sons of Darkness, who comprised the majority of the human race, nor was revenge allowed to individuals (1QS VII,9). The solution was to turn the final decision into an eschatological battle where divine and angelic hosts, led by archangel Michael, would secure victory for the Sons of Light. This war was to be more than its models, the war against Gog of Magog on the mountains of Israel (Ez 38-39) or the battle of the king of south against the king of north in Dan 11:40-12:3. The enemies were to be destroyed once and for all, as according to the traditions of the holy war they were a sacrifice (Mrx) to God (1QM XVIII,4-5).

For the members of the Qumran community violence was only allowed in connection with the eschatological war. But another kind of violence typical of apocalyptic imagination even outside of Qumran frequently appears in their writings.

Apocalypticism is a feature of apocalyptic literature. Neither the War Scroll nor the pesharim nor the Community Rule (1QS) are apocalypses. When I use the term "apocalyptic imagination" here I refer to imagination typical of the apocalyptic era (c. 200 B.C.E. - 200 C.E.). Defined in this way, apocalyptic imagination may also appear outside the genre of the apocalypse. The period when apocalypticism flourished coincided with the time of the Qumran community. Thus it is reasonable to investigate what kind of apocalyptic thinking occurs in the "sectarian" texts of Qumran. The War Scroll, the Community Rule and the pesharim are here examples of "sectarian" literature. It was characteristic of apocalypticism to suggest that this age was approaching its end. Following the end of the age the dawn of a new better age was to be expected. At the turn of the era a period of great distress and cruel suffering was due to occur. This idea is most clearly expressed in 1QS, where the terminus "time of visitation" occurs (1QS III,18; IV,18-19.26). In the time of visitation the faithfulness of the people of Yahweh would be tested (1QM XVII,8-9). In the new world there would be room only for the people of God, the chosen ones, the faithful ones, while the wicked men would be completely destroyed. The faithful ones would not be able to avoid the hardships of the final age; quite the opposite, it was a hard time for them, too, but the interpretation of the community took utmost suffering as indicative of the sufferers' being elected by God and certain of access to the new age.

The hardships of the final age provide an interesting angle on the relation of the Qumran community to force and violence. Suffering, including unjust suffering, was the lot of the elect. Apocalyptic imagination concerning these hardships was full of violence and the use of force, but it is never permissible for the faithful ones to participate in violent activities or to resort to force (cf. 1QS VII,9). They have to suffer and wait until God intervenes. It is one of the principal messages of the Habakkuk Pesher that individual people need to be patient and wait for Yahweh's intervention and the Day of Judgment. The faithful will be saved if they remain faithful (1QpHab VIII,1-2), even though enemies surround the righteous and the final age is delayed (1QpHab VII,7-13). The dualism is as pointed as in the War Scroll and the Community Rule. The Habakkuk Pesher places foreign nations and traitors in the nation of Israel on the same side; both will be punished by God (II,1-6, V,3-5). The pesherist describes "the men of violence" as those "who had rebelled against God" (1QpHab VIII,11). This closely resembles the dualism of the War Scroll and clearly indicates that the community regarded violence as wickedness. The Habakkuk Pesher states that "on the Day of Judgment God will destroy from the earth all idololatrious and wicked men" (1QpHab XII). Similarly, the War Scroll declares that the fate of the Sons of Darkness is defeat. According to 1QS, all those who walk in the wicked spirit will incure severe sufferings until they are totally defeated (1QS IV, 12-14).

Apocalyptic imagination is a kind of sublime form of revenge which is entrusted into the hands of Yahweh. It could be called "eschatological vengeance". There are no trace of a furious "zeal of Phinehas", even though this righteous zeal was approved by Moses in Num 25:10-15 and the writer of 1 Maccabees described it in an idealistic light. The description of the horrors of the final age is at times cruel and dreadful, but it does not exceed the sufferings of real life in wartime. The description is mostly dry and without pathos. I have noticed only few occasions where the vengeance is so vivid that it apparently arose from personal emotions. One example is the passage where the Wicked Priest, who pursued the Teacher of Righteousness, is threatened with the anger of Yahweh (1QpHab XI,1-14). Even though the emotions are involved in the presentation of the fate of the Wicked Priest, the vengeance is Yahweh's. Because his fate was already history, it could be used as an encouraging example of the righteousness of Yahweh and the divine intervention that the community had experienced earlier. Except for the final eschatological battle, the use of force was not an ideal taught to the members of the Qumran Community.

In the apocalyptic imagination the emphasis often lies on the negative side, on descriptions of the sufferings and horrors of the fates of the traitors or wicked people (e.g. 1QS IV, 12-14). The positive side is, however, also recounted (1QS IV, 6-8). The spirits of the community should be raised by dreams of a happy and peaceful time of salvation.

What do we know about the final battle and circumstances at Qumran when the Roman legions approached Qumran on their way to Masada? We know that the members of the community hid their precious scrolls in caves before the legions reached Qumran. But did the members of the community attempt to fight the Romans (the kittim) or did they hide in the caves, too, or did they capitulate without resorting to force? In his Jewish War II, 149-153, Josephus tells that Essenes were terribly tortured by the Romans, but they did not blaspheme their lawgiver or eat any forbidden thing. Josephus' description resembles of Maccabean legends, as they appear in 2 Macc and 4 Macc. It seems likely that his story is not totally trustworthy, but he made the Essenes martyrs and heroes of their faith in the Maccabean spirit. Nevertheless, Josephus does not maintain that they refused the use of arms. On the contrary, he relates that they carried arms on their journeys to protect themselves against brigands (Jewish War II, 125). One interesting detail in this connection is that Josephus mentions in Jewish War II, 567 that one of the Jewish military leaders who were appointed at the beginning of the revolt against Rome in 66 C.E. was called John the Essene. Thus, Josephus does not "explicitly describe the Essenes as pacifists". Josephus speaks of the Essenes in general, but the Qumran Community hardly represented totally pacifistic ideology, either.

That the Romans destroyed the main building of the community points to at least some skirmishes at Qumran. Of course, the members of the community, were destined to loose the fight. Whatever torture or hardships they had to suffer after the battle as prisoners of the Romans, they possibly interpreted them as indicative of the final age when they had to prove faithful.

n Qumran's two Messiahs

The term "moshiach" (Messiah) literally means "the anointed one," and refers to the ancient practice of anointing kings with oil when they took the throne. The moshiach is the one who will be anointed as king in the End of Days. The word "moshiach" does not mean "savior." The notion of an innocent, divine or semi-divine being who will sacrifice himself to save us from the consequences of our own sins is a purely Christian concept that has no basis in Jewish thought. Unfortunately, this Christian concept has become so deeply ingrained in the English word "messiah" that this English word can no longer be used to refer to the Jewish concept. The word "moshiach" will be used throughout this page.

The moshiach will be a great political leader descended from King David (Jeremiah 23:5). The moshiach is often referred to as "moshiach ben David" (moshiach, son of David). He will be well-versed in Jewish law, and observant of its commandments (Isaiah 11:2-5). He will be a charismatic leader, inspiring others to follow his example. He will be a great military leader, who will win battles for Israel. He will be a great judge, who makes righteous decisions (Jeremiah 33:15). But above all, he will be a human being, not a god, demi-god or other supernatural being.

There is not one, single messianology to be found in the texts from Qumran. Instead, we must accept that there are several theories about the Messiah. In the *War scroll* the Messiah is a prophet and takes no part in the war between the 'children of light' against the 'children of darkness', although the Messiah can be identified with the 'prince of the community'. In other texts, the Messiah is a war leader (e.g., 4QFlorilegium and 4Q458). These are clearly conflicting messianologies.

Several texts are considered to be written by members of the sect: the *Damascus document* for example, and the *Messianic rule*. In these texts, we may expect to find the sect's own messianology. The distinguishing characteristic is that the Qumranites expected the coming of not one, but two Messiahs. This must have been an attempt to make sense of such contradictory messianic images as we have encountered up till now.

The root of this idea may be the lines of Zechariah:

> 'Here is the man whose name is the Branch, and he will branch out from his place and build the temple of the Lord. It is he who will build the temple of the Lord, and he will be clothed with majesty and will sit and rule on his throne. And he will be a priest on his throne. And there will be harmony between the two.'

The Dead Sea Jesus

> [*Zechariah* 6.12-13]

This refers to prince Zerubbabel and the high-priest Joshua, but it is certain that it was understood in a messianic sense in the early Hasmonaean period. For example, the author of the *Testaments of the twelve patriarchs* expected a priestly and a kingly ruler:

> My children, be obedient to Levi and to Judah. Do not exalt yourselves about these two tribes because from them will arise the Savior from God. For the Lord will raise up from Levi someone as a high-priest and from Judah someone as king. He will save all the gentiles and the tribe of Israel.
>
> [*Testament of Simeon* 7.1-2]

> To me [=Judah], God has given the kingship, and to him [=Levi], the priesthood. And He has subjected the kingship to the priesthood. To me He gave earthly matters and to Levi heavenly matters. As heaven is superior to the earth, so is God's priesthood superior to the kingdom on earth.
>
> [*Testament of Judah* 21.2-4a]

The word 'Messiah' is not used, however. This step was taken by the sect at Qumran. Its members were looking forward to a 'Messiah of Israel' and a 'Messiah of David', who resemble the kingly and priestly descendants of Judah and Levi in the *Testaments of the twelve patriarchs*.

The first text we must study is the Damascus document, which is, as so often at Qumran, a combination of texts. Its first part is a kind of theological history which proves that the sect is the true Israel and that God will reward the faithful; then follows a kind of law; and a brief penal code is added as an appendix. Our first quote does not mention the Messiah, but must without any doubt be interpreted in a messianic fashion, because it alludes to Balaam's prophecy.

> And the star is the seeker of the law, who came to Damascus; as it is written *A star has journeyed out of Jacob and a scepter is risen out of Israel*. The scepter is the Prince of the whole congregation, and at his coming *he will break down all the sons of Seth*.
>
> [*Damascus document* 7.18-21]

This is a very interesting text, because it not only mentions two Messiahs, but also shows that one of them is a military leader and the other a sage. Moreover, the expression 'seeker of the law' usually signifies the Teacher of righteousness (the founder of the sect); the fact that this title is now used to describe one of the Messiahs suggests that the members of the Qumran sect believed that he would one day return.

In this first quote, the word 'Messiah' is not used. But the *Damascus document* is sometimes more explicit.

> [...] during the time of ungodliness until the appearance of the Messiahs of Aaron and Israel [...]
>
> [*Damascus document* 12.23-13.1]

> This is the exact statement of the ordinances in which they walk until the Messiah of Aaron and Israel appears and expiates their iniquity.
>
> [*Damascus document* 14.18-19]

> Those who heed Him are the poor of the flock; they will be saved at the time of visitation. But others will be delivered up to the sword at the coming of the Messiah of Aaron and Israel.
>
> [*Damascus document* 19.9-11]

(Similar ideas can be found in 19.33-20.1)

At least one text adds a third actor in the messianic age: the prophet. He is mentioned in the *Manual of discipline*. This text is also interesting because it uses the plural 'Messiahs of Aaron and Israel' instead of the singular 'Messiah of Aaron and Israel' of the *Damascus document*: this proves that there were indeed two (or three) Messiahs.

> And they shall not depart from any counsel of the law to walk in all the stubbornness of their heart, but they shall be governed by the first ordinances in which the members of the community began their instruction, until the coming of the prophet and the anointed ones of Aaron and Israel

[*Manual of discipline* 9.9b-11]

Just like the kingly Messiah of Israel and the priestly Messiah of Aaron, the prophet is a messianic type, and it is possible to believe that the Qumran library also contained a messianology that assumed that there would be *three* Messiahs. After all, kings, priests and prophets were the only one that could be anointed.

The most interesting text, however, can be found in the Messianic rule (also called Rule of the congregation). It describes the table arrangement during a sacred, messianic meal. The interesting point is the hierarchy between the two Messiahs.

> This is the sitting of the men of renown called to the assembly for the council of the community when God will have begotten the Messiah among them. The Priest shall enter at the head of all the congregation of Israel, then are all the chiefs of the sons of Aaron, the priests, called to the assembly, men of renown. And they shall sit before him, each according to his rank.
> Afterwards, the Messiah of Israel shall enter. The chiefs of the tribes of Israel shall sit before him, each according to his rank, according to their position in the camps and during their marches; then all the heads of family of the congregation, together with the wise men of the congregation, shall sit before them, each according to his rank.
> And when they gather for the community table, or to drink wine, and arrange the community table and mix the wine to drink, let no man stretch out his hand over the first-fruits of bread and wine before the Priest. For it is he who shall bless the first-fruits of bread and wine, and shall first stretch out his hand over the bread. And afterwards, the Messiah of Israel shall stretch out his hands over the bread. And afterwards, all the congregation of the community shall bless, each according to his rank.

[1Q28a 2.11-21]

Probably, the idea of a dual Messiahship did not disappear with the sect (which came to an end during the war between the Jews and Romans of 66-70). It is possible, but certainly not proven, that Simon ben Kosiba was recognized as the Messiah of Israel and his collaborator Eleazar of Mode'in as the Messiah of Aaron.

6. Archeology of Qumran

a. Dating the findings

There are basically three methods used in dating the Dead Sea Scrolls: paleography, Radio Carbon Dating (including AMS), and archaeology. The early paleographic work done on the scrolls was done by Frank Cross, and he and his team concluded that the scrolls were dated between the Maccabean and Roman periods, with different scrolls being dated to specific periods within this frame of reference. Radio-carbon dating and AMS have basically confirmed the paleographic dating of the scrolls. The archaeological work done by people like Roland de Vaux and Jodi Magness at Qumran has also shown all of the paleographic and radio-carbon dates to be plausible dates for the production and/or collection of the scrolls by the Qumran community. Major documents like CD and IQS can be dated to around 100 CE. Hypotheses that make the Qumranites and their scrolls Christian (Robert Eisenman, Barbara Thiering) have to reject these paleographic, radio-carbon/AMS, and archeological conclusion concerning the dating of these documents.

b. Physic´ Space of the Community: Historical Evolution

At the end of the First Temple period (8th-7th centuries BCE), a first settlement was established at the site. Sparse remains of a small, fortified farmhouse or Judahite fort were found. The site was identified by some as Secacah, or the City of Salt, two of the six cities in the desert territory of Judah. (Joshua 15:61-62)

Settlement at Qumran was renewed at the end of the 2nd century BCE, probably during the reign of the Hasmonean King John Hyrcanus I, when the existing structure was restored and enlarged. Then, at the beginning of the 1st century BCE, during the reign of Alexander Jannaeus, renewed building determined the plan of the site until its destruction. An aqueduct was built from a cliff above Wadi Qumran several hundred meters east of the site. Winter floodwaters were collected behind a dam at the foot of the cliff and from there flowed in the aqueduct to Qumran and filled the numerous cisterns and mikva'ot (ritual baths) there. The supply of water was essential to a permanent settlement at Qumran, where summer temperatures in this desert region are extremely high.

The plan of Qumran is unique, not at all similar to other contemporary settlements, with its many large halls, undoubtedly serving public functions, and the relatively small number of living quarters. The main entrance to the settlement was in the north, at the foot of a watchtower. The walls of the buildings were made of stones gathered at the foot of the cliff and plastered with thick, white-gray plaster. The windows and doorposts were built of well-trimmed stones and the roofs, as was common in that period, were constructed of wooden beams, straw and plaster.

The main structure at Qumran had several rooms, some obviously two stories high, arranged around a central courtyard. In the northwestern corner was a square watchtower with particularly thick walls that rose above the rest of the settlement. The tower served as a lookout and warning post and protected the settlement against raids by desert tribes. A room with benches built along its walls served as a meeting-place for the members of the community and probably as a place for Torah (Bible) study. Additional building complexes, south and east of the main building contained long halls, rooms and ritual baths. One of the large halls was for meetings and served as a refectory. In a storage room and a kitchen next to it, neat piles of hundreds of pottery vessels and a large number of small food bowls were found. A workshop, in which pottery vessels for use of the community were produced, was discovered in the southeastern part of the site. The workshop included a basin for preparing the clay, a potters wheel made of stone and two round kilns for firing.

A large number of mikva'ot (ritual baths) was found throughout the site. Excavated into the marl soil, they were waterproofed with thick, gray hydraulic plaster. The broad staircase leading to the bottom was at times divided down the middle by a low (20 cm.-high) wall, which separated those descending for immersion from those leaving after purification. The ritual baths were fed by water from the aqueduct. Mikva'ot similar to those at Qumran were typical of public and private buildings in Jerusalem and elsewhere in the Second Temple period. The Mishnah (Masekhet Mikva'ot) notes the importance of immersion in water for spiritual purification and lists the requirements for such ritual baths. The mikva'ot at Qumran were built according to all these requirements. Unusual at Qumran is the large number of these installations and the size of some of them, relative to the settlement. The latter probably served the members of the community for communal immersion, a central part in their daily rituals.

An earthquake severely damaged the buildings and mikva'ot of Qumran in 31 BCE. Excavations revealed cracks in walls and a thick layer of ash from a fire that had raged. The earthquake was mentioned by Josephus. (Antiquities 15, 121 ff.; Wars I, 370 ff.)

The settlement at Qumran was subsequently abandoned, until the beginning of the 1st century CE, when members of the community returned and settled there once more. They restored the earlier structures and, with various additions and modifications, used them. In the main building was a long room, in which remains of benches, or low tables, made of mud and plastered on the outside, as well as small clay inkwells were found. According to the excavator, these finds indicate that the room was a scriptorium, where the settlement's scribes copied the holy writings and the laws governing the community.

Perhaps only a few dozen of the leaders of the community lived permanently at Qumran. Most of the members of the sect, probably totaling several thousand, lived in villages and cities. A large Essene community certainly lived in Jerusalem (according to Josephus, the name of the gate in the southern wall of Jerusalem, at

Mt. Zion, was called the Essenes' Gate). For certain periods of time, members of the sect lived in the desert near Qumran and during holidays and community events, many more arrived and lived in tents, huts and the caves nearby. In a survey and excavations conducted recently in the caves on the marl slopes north of the site, pottery vessels were found, indicating the use of the caves as dwellings. Stone circles nearby also indicate a tent encampment.

The buildings at Qumran were blocked in the east by a wall of large stones. Beyond it, the marl terraces extend several hundred meters before ending in a cliff. On this marl surface was a large cemetery with over 1,000 graves in north-south oriented rows. A few were excavated, revealing the simplest of individual graves dug into the marl and covered with a pile of stones. Most of those buried weremales, though at the edge of the cemetery there were also graves of females and children. The settlement at Qumran was destroyed during the Jewish War against Rome in 68 CE, and it was never resettled.

c. The Manuscripts of the Complex

The state of preservation of the manuscripts varies from almost complete to almost non-existent. Many of the manuscripts are made up of more than one fragment. Once the fragments had been reassembled into manuscripts, each manuscript was given the series designation provided in this list. Some manuscripts consist of a single small fragment. Others contain nearly the entire text of the original. Many unclassified fragments remain unidentified; neither a part of one of the larger scrolls nor a part of any known text. These fragments were each assigned their own unique series designations.

In general, the task has been to assemble smaller fragments into larger fragments wherever joins can be identified. The jigsaw puzzle aspect of the assembly ends when all the available joins have been identified. The assembled larger fragments have typically been collected together based on other evidence into larger manuscripts even when no common joins existed. Such evidence includes language, letter shapes, spacing between rows and columns, widths of the columns, the color of the scroll material, the nature of the damage to the scroll material, the nature of the text [especially if it is from a known text for which a more complete version exists, etc.] Some of this is straightforward and some of it is not. There is the chance that future scholarship will force a revision in some of these assignments, however, everyone seems to agree that the job done to date was done very thoroughly and with a high degree of care, skill and precision.

Some of these manuscripts are copies of the same, or nearly the same text. Each manuscript copy received its own distinct series designation. But for many of these copies the same official abbreviations and/or names are often used. To distinguish among the copies, superscript letters are often used when referring to them by name.

Most of the early manuscripts and a high percentage of the Cave 4 manuscripts were not acquired through personal excavation by the official archaeological expeditions. They were purchased from the Bedouin who found them. The buyers were primarily the representatives of Jordan and Israel. This makes it impossible to assign specific fragments and documents to specific caves with complete confidence (chain of custody and provenance are undocumented). It is not even entirely certain that all manuscripts discovered by the Bedouin have been accounted for. Comments about the distribution of documents among the various caves and discussions of why certain manuscripts were stored in certain caves must include the implicit proviso that it is all subject to change should more data or manuscripts become available. (Note that finding a fragment of a purchased manuscript in one cave does not necessarily prove, only improves the likelihood, that the purchased manuscript was originally taken from that same cave.)

Initially, de Vaux and Milik divided the texts into biblical (included in the Hebrew Bible) and non-biblical categories. The term "non-biblical" should not be understood as non-religious. Almost all the works in the Qumran library are religious in some sense. "Non-biblical" simply means not currently part of the accepted Jewish Canon. In other words, these are among the texts that did not make it into the Bible.

Over time the editors have occasionally chosen to renumber and rename certain of the manuscripts. This seems to have been due in part to their evolving understanding of how the fragments and manuscripts fit together. Furthermore, not all scholars who have studied the texts agree on how each of them should be reassembled from the available fragments. For these, and perhaps other, reasons, there are occasional missing numbers.

It is important to remember that these series designations are intended to refer to individual manuscripts. There are many techniques that can be used to determine if two fragments of one text are from the same or separate manuscripts. These include the color and texture of the parchment or papyrus on which it is written, and the handwriting, language and idiomatic usages of the scribe(s) who wrote it (them). It should be obvious that if even a single part of the two fragments overlap, then two separate manuscripts are, almost certainly, required. On the other hand, many fragments with no overlaps and no contiguous edges with the other fragments, have been assigned to specific larger documents. The techniques used in making these assignments are not infallible, and it is always possible that future scholarship and/or investigative techniques will require reassignment of some fragments.

Manuscripts or fragments, now numbered separately, may turn out to be parts of other numbered manuscripts. While most of the details of this jigsaw puzzle were worked out long ago, it is still possible that some of the unidentified individual fragments, currently carrying their own unique manuscript designations, may yet be identified and, possibly, incorporated into other manuscripts. This would possibly create additional gaps in the series numbering. It is also possible that a fragment now assigned to one document might turn out to be part of another copy of the same text or even part of an unrelated text. Such a fragment could, in the latter case, require its own new number.

Manuscripts with non-numerical official designations (such as the first seven manuscripts) appear at the beginning of the list for the appropriate cave (Cave 1 for those first seven manuscripts). Some famous or notorious manuscripts have become better known by their official abbreviations or one of the common names than by their numerical series designation.

In a few special cases, one manuscript consumes two numerical series designations. This occurred because parts of the manuscript ended up in Israel and part of it ended up in the Rockefeller Museum basement in East Jerusalem. Given the temper of the times and of some of the individuals involved, there was no way to reunite the separate parts. Today, it should be possible, but there are no signs that such reunions have actually occurred under the auspices of the Israel Antiquities Authority.

Until recently most of the non-biblical texts have been only partially published or not published at all. These texts are potentially more interesting than the biblical texts, in part, because they are among the lost religious texts of the intertestamental period. What is even more interesting, they were lost without leaving us any trace that they ever existed; at least, not until the late 1940's. As the Damascus Document discoveries in the Cairo Genizah demonstrate, however, some of these may have been lost more recently than might be suspected. Still, it is always most interesting to stumble across the totally unexpected. The newly won availability of these texts now offers scholars an opportunity to start digging for the surprises.

8. Judaism in Qumran

As Judaism emerged from the Babylonian exile (539 B.C.), it inherited the stress of Israelite religion on monotheism: "Hear, O Israel, the Lord our God, the Lord is one . . ." (Deut 6:4).

a. Temple and Priesthood

The first Temple was built by David's son, Solomon, in the tenth century B.C.E. and destroyed by the Babylonians in 587 B.C.E.. A modest Temple was rebuilt by the returning exiles in 515 B.C.E. and further reconstructed on a grand scale in the Roman-Herodian period. This reconstruction was begun by Herod the Great in 20 B.C.E. and was not completed until about 60 C.E., only to be destroyed a decade later. In the Persian period, the priests gained power due to the absence of an actual king and the decline of prophecy; in fact, the High Priest, as leader of the cultus and interpreter of religious traditions became the most powerful figure in Judaism. Under the Seleucid Greeks, the High Priesthood became something of a political position; then the Maccabees (who were also of priestly descent, though of an undistinguished line) assumed control of the High Priesthood and eventually assumed royal prerogatives as well, thus succumbing to the politicization of the office. Hence, other priestly parties made their appearance, among them the Essenes and the Sadducees. Under the Herodians and procurators, High Priests were of varying families, and they were appointed to the post; nonetheless, they maintained a measure of political power, for they continued to preside over the central cultus at the Temple and over the religious Sanhedrin, Judaism's highest court. The destruction of Jerusalem and the Temple in 70 C.E. meant the end of their power.

Apart from the political functions of the priests, their major religious functions consisted of the maintenance of purity by the sacrificial system at the Temple. In Judaism, sin was not only a moral question; it also concerned the practice of ritual and notions of the sacred and profane, purity and impurity--distinctions that are often lost to the modern consciousness. In ancient Israel a whole system of sacrifices had arisen to atone for sin, that is, to set sinful humankind right with the one, holy God. The priests administered the system and sacrifices were offered at least twice a day. Even the architectural plans of the successive temples reflect the various degrees of holiness. For example, only the outermost area of the Herodian Temple was accessible to Gentiles; beyond it they could not go "under penalty of death." Moving toward the center in the Sacred Enclosure (for Jews) was the Court of Women, the Court of Israel (men), the Court of Priests, and the Holy Place--the forecourt where the sacrifices took place, and finally the Holy of Holies into which the High Priest entered only once a year, on the Day of Atonement. Thus the Temple was the holy center of the holy city in a holy land. Yet, like all oriental temples, it was also the hub of much economic and commercial activity, for it housed the national treasury. Every Jew was expected to pay the annual Temple tax.

b. Synagogue and Prayer

Sacrifice was an enacted prayer, that is, a means of human communication with God. There were also other forms of liturgical prayer; for example, the whole tradition of chants and psalms which in New Testament times had become the special province of a class of Temple priests, the Levites. This form of public prayer was continued even where there was no access to the Jerusalem Temple. When the synagogue (from the Greek for "gathering together") developed sometime in the post-exilic period (the earliest archeological evidence is from the first century C.E.), it served as a "house of prayer" as well as a gathering place for meetings, meditation, and instruction. No sacrifice was offered there. Rather, the synagogue services probably consisted of a recitation of the *Shema* ("Hear, O Israel, the Lord our God, the Lord is one . . ."), Scripture, sermon, blessing, and, of course, prayer. Prayer0s could be offered at any time and any place; yet, they should be oriented toward Jerusalem--specifically the Holy of Holies--and it was customary to offer them three special times a day, namely, morning, midday, and evening. Standing or kneeling with hands raised to heaven were the usual positions.

c. The Centrality of Torah

In the post-exilic period, Judaism sought God's will more and more in sacred tradition and the written word and its interpretation became the very basis of life. *Torah* meant "instruction": in its widest sense any form of revelation; in a somewhat narrower sense, Scripture and its written and (especially) oral interpretation; and in a still narrower sense the Pentateuch (Five Books of Moses)--most specifically the legal materials in the Pentateuch. It was therefore "law" but included narrative materials as well. For a summary of its main elements see the document on this Web page: Summary of the Torah.

d. Apocalyptic Eschatology

The term eschatology is from the Greek terms *eschaton*, "the end," and *ho logos*, "the word," "the teaching." It means therefore, "teaching concerning the end of things"--specifically, teaching concerning the end of the world. A particular form of eschatology is called "apocalyptic" (from the Greek *apocalypsis*, "an uncovering," "a revelation,"); it describes both a movement and a literature that characteristically claimed that God had revealed to a writer the secrets of the imminent end of the world and so had given him a message for his people. As with Wisdom, the literature dates after 200 B.C.E. and is largely non-Biblical (that is, outside the Old Testament). It reveals a very diversified Judaism prior to 70 C.E., one marked by a number of movements which, if measured by the Judaism that survived the wars, appears in many respects non-normative or unusual. Much of that literature is the literature of apocalyptic eschatology.

There is no absolute agreement about what constitutes apocalyptic eschatology either with respect to its origins or content. It shows influences of Old Testament prophecy and Wisdom literature; but there are also currents of Persian dualism and Babylonian astrology. It is a child of hope and despair: hope in the invincible power of God, the world he created, and his plan and purpose for his people, but despair over the present course of human history in that world. The primary tenent of Jewish faith was that one true God was the creator and the ruler of all within it. At the same time, the actual experience of the people of God in the world was catastrophic: Assyrian and Babylonian conquest, exile in foreign lands, Persian domination, the coming of the Greeks, and finally the Romans. The burdens of war, occupation, forced Hellenization, and taxation by imperialistic powers produced an intolerable experience of alienation and powerlessness. Human history was a virtual descent into hell. But God was the ruler of all things and, therefore, he must have foreordained the tragic events of human history. Thus, there was some divine plan through which the horrors of history would reach a climax and everything would change. The hope was that the world would become much the same as it had been in the beginning of time: a paradise in which God's elect people would be vindicated. This change would be marked by tremendous historical and cosmic catastrophes. In the meantime the people of God had to prepare themselves for the change and watch for the signs of its coming.

The most apocalyptic book in the Old Testament is the book of Daniel, which contains the Son of Man vision in 7:13-14, highly influential in the gospels:

> *I saw in the night visions,*
> *and behold, with the clouds of heaven*
> *there came one like a son of man,*
> *and he came to the Ancient of Days*
> *and was presented before him.*
> *And to him was given dominion*
> *and glory and kingdom,*
> *that all peoples, nations, and languages*
> *should serve him;*
> *his dominion is an everlasting dominion,*
> *which shall not pass away,*
> *and his kingdom one*
> *that shall not be destroyed.*

There are many other forms of the apocalyptic hope. The *Assumption of Moses*, a work contemporary with the New Testament, is particularly interesting because of its use of "Kingdom of God," a key concept in the teaching of Jesus. Another form of this hope is associated with the Coming of a Son of David found in the first century B.C.E. document called the *Psalms of Solomon*. Despite the variety of the forms of expression, the hope for a climactic series of events that will lead to the final, eschatological intervention of God into human history, directly or through intermediary figures is constant. Through these events the world would be forever changed, transformed into a perfect world in which the people of God would be forever blessed for their fidelity, and their enemies and God's forever punished.

This hope is called the "apocalyptic" hope because the characteristic claim of the literature that expresses it is that God has uncovered or revealed to the writer or seer his plan for the further course of history and the coming of the End. This revelation frequently takes the form of dreams or visions, which are then interpreted by a heavenly figure. The dreams or visions generally use symbols to recount the history of the Jewish (or Christian) people and to express the hope for the immediate future. So, for example, Daniel 7 tells in symbols the history of the Near Eastern world from the Babylonian Empire through the Persian Empire to the conquests of Alexander the Great and his ten successors as kings of the Macedonian Seleucid Kingdom of Syria. The final symbol used to represent a king is the "little horn" (Dan 7:8), which represents Antiochus IV Epiphanes, who began persecuting the Jews in 167 B.C.E. in an attempt to consolidate his empire. The result was the Jewish revolt. The author of Daniel 7 is living at the time of this Maccabean revolt, writing to inspire his people with confidence that the war is the beginning of the End, that it will shortly be ended by the coming of the Son of Man as judge and ruler of the world.

The book of Daniel is pseudonymous, that is, it was written under an assumed name long after the time of most of the events it pretends to prophesy. This is characteristic of Jewish apocalyptic writings, and usually a name of some importance--Abraham, Moses, David, or the like--would be chosen. This feature, of course, lent the writing a certain authority and there was no modern notion of fraud or copyright. The history would be portrayed in symbolic form leading up to the symbolic vision of the seer. The seer also dreamed and thought in traditional symbolic images, and frequently he alluded to previously written texts that contained them.

These are the most important characteristics of apocalyptic eschatology: a sense of alienation and of despair about history that bred the belief that the world was rushing to a foreordained tragic climax, a hope in God that fostered the conviction that he would act in the climactic moment to change things utterly and forever, and a conviction that it would be possible to recognize the signs of the coming of that climactic moment. Its chief literary characteristics were pseudonymity, symbolism, and quotation from previously existing texts.

Associated with some apocalyptic eschatological texts is the hope for a future redeemer, a Messiah. Originally the term "Messiah" (Hebrew *mashiach*; Greek *Christos*) meant "anointed;" in the Old Testament it was applied to any figure that was installed into office by anointing, that is, prophets, priests, and kings. Any of these figures was an "anointed one" or messiah. In the eschatological materials, there are several types of expectation. We have just noted a future redeemer and judge, the Son of Man. Other Jews hoped for a descendant of David to come, overthrow the enemies, and reestablish the Davidic kingdom. In the Dead Sea Scrolls, there is evidence for a triple expectation: a prophet like Moses, a royal Messiah of the line of David ("the Messiah of Israel"), and a priestly Messiah ("the Messiah of Aaron"). The following passage combines this with adherence to the Torah:

> And they shall not depart from any maxim of the Law
> to walk in all the stubbornness of their heart.
> And they shall be governed by the first ordinances
> in which the members of the Community began their instruction,
> until the coming of the Prophet and the Anointed (Ones) of Aaron and Israel.
> *The Community Rule* 9:9-11

8. Was Christ in Qumran?

a. Introduction

The Dead Sea Scrolls enhance our knowledge of both Judaism and Christianity. They represent a non-rabbinic form of Judaism and provide a wealth of comparative material for New Testament scholars, including many important parallels to the Jesus movement. They show Christianity to be rooted in Judaism and have been called the evolutionary link between the two.

Jesus may have visited Qumran, or even lived there for a time, but there is no evidence that he did. He was, however, certainly in the vicinity. He was baptized in the Jordan River (Matthew 3:13), a bare three miles away.

After his baptism, Jesus was "led by the Spirit into the wilderness," where he was "tempted by the devil" (Matthew 4:1; Mark 1:12, Luke 4:1-2). The traditional site of the Mount of Temptation is seven miles north of Qumran, Jerusalem itself is only 13 miles to the west.

That Jesus may have had some personal association with Qumran, given the similarities between the Dead Sea Scrolls and Christianity, is, however, pure speculation. On the other hand, a reasonable case can be made that John the Baptist lived, at least for a time, at Qumran. Early in his life, John lived in the Judean desert; according to Luke 1:80, "The child grew and became strong in spirit, and he was in the wilderness till the day he appeared publicly to Israel."

b. Historical Jesus and the Community of the Yahad

Some see Christianity borrowing from concepts accepted by the Essenes and illustrated in the literature at Qumran. There is, however, very little evidence to substantiate this supposition.

Their major similarities stem from the fact that both Christian and Essene philosophies keyed heavily on prophecy. This approach has major theological implications and accounts for many of the similarities between each of their literatures.

But there is a critical difference in their interpretations. The sectarians of Qumran did not see the prophets modifying the Torah, an essential tenant of Christian thought. They ascribed a descending order of importance to the Hebrew scriptures and it was a hierarchy determined entirely by age. The older the document, the more credence it carried.

The Essenes were quite visible at the time of Christ and all the people knew them; they were scattered throughout Palestine, in every village. Yet Christ never found it necessary to single them out as objects of falsehood the way he did the Pharisees or Sadducees.

In fact, if the works prescribed by the Jewish Law could save, the Essenes would have been the proof of it. The fact is, Christ prescribed a completely different theological formula for God's salvation; one based on faith and coming outside the Law.

The Jewish philosophers and historians may have considered the Essenes the paragons of virtue in the Palestinian theocracy. The Dead Sea Scrolls show us that the Essene's were exclusive and secret; they believed in a kind of Gnosticism: a secret knowledge known only to privileged insiders. They hated women, were intolerant of children, frowned on marriage, forgave each other but hated their sworn enemies, and practiced rigorous fidelity to infinite refinements of the physical Law.

They were fanatical in their ideas of clean and unclean; bathing even before sitting down to meal as a matter of theological ritual. They were afraid even to cook a meal on the sabbath. Yet they fed the poor and helped widows and orphans, wedding the themes of the prophets to the restrictions of the Torah.

A major focus of Qumran literature is their discipleship with the prophets and their obsession with prophecy. This contrasts markedly with the focus of the Rabbi's of orthodox Judaism.

The Essene idea of an immortal soul was only a modest refinement of the pharisaitic philosophy expressed in the writings of Josephus, and both ideas stemmed originally from the Greek philosopher Plato who introduced the doctrine some 400 years earlier.

The Essenes can be seen in two contexts. One, they were convinced that they could work their way into God's favor and had developed one of the most sophisticated Jewish systems for such an endeavor up to that time.

Numerically they were not that successful, gaining only about 4000 adherents over a period of at least two centuries. Their small numbers in relation to the other people of Palestine at the time of Christ were almost certainly the reason why no mention of them is found in Christian literature.

Their's was only one of a chorus of voices that heralded the coming of the Messiah to Palestine at the time of Jesus. This time in the history of Palestine was called the 'Messianic Age' because of the fervor of the people in expectation of the fulfillment of the prophetic writings.

Their teaching was a counterbalance to the philosophy of the Pharisees and Sadducees; refocusing attention on the social themes of the prophets at a time when justice was becoming a theological orphan.

Although these were voices of prophecy, and a part of the Messianic fervor of the times, the Essenes were not to be the fulfillment of the prophecy that they heralded.

The Essene theology looked forward to the coming of the kingdom. In the Qumran writings the Teacher of Righteousness was not considered the Messiah; he was a herald who taught that the Messiah's coming was imminent, and that preparations needed to be made for the sake of being able to stand before that appearance in a state of worthiness.

The members of the sect of Qumran cherished messianic expectations. We know from the New Testament that such anticipations were widespread in Judea during this period. And in the 2nd and 1st centuries B.C., a number of works were written in which these expectations were expressed.

In the New Testament we have only one messiah expected: the descendant of David. In the Scrolls at Qumran, however, we find the expectation of two Messiah's; one Davidic and the other Aaronic. In other words, one a king and the other a priest.

In the Manual of Discipline we find the expression "the messiahs of Aaron and Israel" used, i.e., a Messianic Pair. Greater status, in fact, at Qumran was given to the priestly Messiah -- who was termed the 'Anointed One'.

Such a preference is not surprising. The Jewish focus has always been on the priesthood. The holiest lineage in Jewish theology in this regard is descendancy in the priesthood of 'Zadok', a term used liberally by the sectarians of Qumran. For them, Zadok defined the Hebrew priesthood, and with it, the core of messianic expectation.

Christians, ignore this high priest of Hebrew history, and concentrate their entire focus on 'David', the source of royalism in the Old Testament, and under whom Zadok co-governed (with Abiathar) the religious ritual of Israel.

Abiathar was defrocked by Solomon, David's son, and from that point on, the sons of Zadok came into complete control of the Hebrew priesthood, forming a ruling aristocracy that continued down through the centuries until the stormy days that surrounded Antiochus Epiphanes IV.

The 'Zadokite Work', found at the beginning of this century in the Cairo Genizah and now generally recognized as having emanated from Qumran, shows that there was an expectation that the two Messiah's would come within 40 years of the death of the Teacher of Righteousness.

For the Christian church Jesus is the only Messiah, and it has no place for a second. He is believed to be the Davidic Messiah, the idea of which is drawn from the Old Testament, not from Qumran; and there is no indication anywhere in Christian writings that a priestly Messiah was ever contemplated, or would have precedence over Jesus.

This issue is so fundamental, and the differences between Christians and the Qumran sect so diverse in this regard, that it makes any thought of Christianity growing out of Qumran doctrine virtually impossible to seriously accept, even if no other differences existed between them.

c. Conclusions

Thus, the scrolls do not have a direct connection with Jesus or early Christianity but they do provide a vastly important context. From them we get a direct glimpse into the world out of which Christianity grew. For those who want to understand the history of Christianity, the scrolls are exciting and enriching. For those who see Christianity and Christian doctrine as something entirely new and unrelated to its Jewish history, the scrolls are threatening.

Some have sought to draw parallels between figures in the scrolls and John the Baptist or Jesus, but an objective examination of such parallels reveals that the differences are greater than the similarities. Any contact of Jesus with Qumran is entirely speculative and most improbable. The suggestion that John the Baptist may have spent some time with the Qumran community is possible, since the Gospels tell us that he spent considerable time in the wilderness near the area where the Qumran community is located (Mt. 3:1-3; Mk. 1:4; Lk. 1:80; 3:2-3). John's message, however, differed markedly from that of the Qumran brotherhood. The only real common point was that they both taught that the "kingdom of God" was coming.

One of the most important contributions of the Dead Sea Scrolls is the numerous Biblical manuscripts which have been discovered. Until those discoveries at Qumran, the oldest manuscripts of the Hebrew Scriptures were copies from the 9th and 10th centuries AD by a group of Jewish scribes called the Massoretes. Now we have manuscripts around a thousand years older than those. The amazing truth is that these manuscripts are almost identical! Here is a strong example of the tender care which the Jewish scribes down through the centuries took in an effort to accurately copy the sacred Scriptures.

9. The Stone that Killed God

a. "Hazon Gabriel" or "The Vision of Gabriel"

Ada Yardeni and Binyamin Elitzur recently published the text of a fascinating text they call "Hazon Gabriel" or the Gabriel Revelation. This text, engraved in stone, conveys the apocalyptic vision of the Archangel Gabriel. Yardeni and Elitzur date it by its linguistic features and the shape of the letters to the end of the first century B.C.E.

The first mention of the "slain Messiah" called Mashiah ben Yosef (Messiah Son of Joseph) is in the Talmud (Sukkah 52a). The story of this slain messiah could be based on historical fact: it is believe to be connected to the Jewish revolt in the Land of Israel following the death of King Herod in 4 B.C.E. This Jewish insurrection was brutally suppressed by the armies of Herod and the Roman emperor Augustus, and the messianic leaders of the revolt were killed. These events set the slain Messiah Son of Joseph tradition into motion and paved the way for the emergence of the concept of "catastrophic messianism." Interpretations of biblical text helped to shape the belief that the death of the messiah was a necessary and indivisible component of salvation. Certain groups believed the messiah would die, be resurrected in three days, and ascend to heaven (see "The Messiah Before Jesus," 27-42).

It seems, though, that this is a stone inscription discovered in the Transjordan, with a text that presents an eschatological revelation by the angel Gabriel and which mentions David, Ephraim, the prince of princes, the slain of Jerusalem, and the merkavah (chariot - God's throne-chariot).

The stone is an Hebrew inscription of 87 lines, written in ink on a large stone. Its precise provenance is unknown. The text is arranged in two columns, similar to the columns in a Torah scroll, and is written in a 'Jewish' script of the late first century BCE resembling the script evidenced in Qumran scrolls; however, its contents and style are different. The text contains a verse from the biblical book of Haggai, with minor changes, and expressions from Zechariah and Daniel.

It also contains expressions from later Jewish literary sources, such as Hechalot literature, Piyyut, Talmud, and Midrash, as well as some that have no parallels elsewhere. Due to its bad condition, the inscription is difficult to interpret, but the expression which may be translated as 'thus said YHWH, the God of armies, the God of Israel' appears many times, with slight variations, similarly to expressions in biblical prophecies, and the name Jerusalem is mentioned several times. The text is written in the first person, the speaker identifying himself as 'I, Gabriel', probably referring to the angel by this name. It seems that the composer of the text belonged to the supporters of the Davidic dynasty and may have been addressing his opponents. However, since no similar text has been discovered to date, it is difficult to determine its precise nature.

b. Polemic Contents

According to Dr. Knohl's reading, an scholar from Israel, lines 80 of the Gabriel text should be read:

"By three days-live, I Gabriel command you, prince of princes, the dung of rocky crevice."

The three day statement is surely fascinating in the light of Jewish views of the afterlife, but even more interesting is that this particular corpse, that Knohl identifies as that of the crowned Jewish rebel leader Simon, killed in Transjordan in the 4 BCE revolt following the death of Herod the Great, is spoken of as "dung" in the rocky crevices where he was slain. Knohl's main point at the conference was that the Jewish idea of "making live the dead" did not necessarily involve the revivification of a copse, as in this case one turned to "dung," but rather a revived life in what would be potentially a "new body."

In 4 BCE, king Herod the Great died. Immediately, there were several revolts against the rule of his son and successor, Herod Archelaus. One of the rebels was Simon of Peraea, who claimed the kingship for himself. The fact that was a slave, is of no importance: slaves could be highly educated and civilized people. The Roman historian Tacitus heard about this incident too, because he writes: "At Herod's death, without waiting for imperial decision, a certain Simon usurped the title of king. He was dealt with by the governor of Syria, Quinctilius Varus, while the Jews were divided up into three kingdoms ruled by Herod's sons." (Tacitus, Histories 5.9.2)

Much of the text, a vision of the apocalypse transmitted by the angel Gabriel, draws on the Old Testament, especially the prophets Daniel, Zechariah and Haggai. Ms. Yardeni, who analyzed the stone along with Binyamin Elitzur, is an expert on Hebrew script, especially of the era of King Herod, who died in 4 B.C. The two of them published a long analysis of the stone more than a year ago in Cathedra, a Hebrew-language quarterly devoted to the history and archaeology of Israel, and said that, based on the shape of the script and the language, the text dated from the late first century B.C.

A chemical examination by Yuval Goren, a professor of archaeology at Tel Aviv University who specializes in the verification of ancient artifacts, has been submitted to a peer-review journal. He declined to give details of his analysis until publication, but he said that he knew of no reason to doubt the stone's authenticity.

Simon who was slain by a commander in the Herodian army, according to the first-century historian Josephus. The writers of the stone's passages were probably Simon's followers, Mr. Knohl contends. The slaying of Simon, or any case of the suffering messiah, is seen as a necessary step toward national salvation, he says, pointing to lines 19 through 21 of the tablet - "In three days you will know that evil

The Dead Sea Jesus

will be defeated by justice" - and other lines that speak of blood and slaughter as pathways to justice.

To make his case about the importance of the stone, Mr. Knohl focuses especially on line 80, which begins clearly with the words "L'shloshet yamin," meaning "in three days." The next word of the line was deemed partially illegible by Ms. Yardeni and Mr. Elitzur, but Mr. Knohl, who is an expert on the language of the Bible and Talmud, says the word is "hayeh," or "live" in the imperative. It has an unusual spelling, but it is one in keeping with the era.

Two more hard-to-read words come later, and Mr. Knohl said he believed that he had deciphered them as well, so that the line reads, "In three days you shall live, I, Gabriel, command you."

To whom is the archangel speaking? The next line says "Sar hasarin," or prince of princes. Since the Book of Daniel, one of the primary sources for the Gabriel text, speaks of Gabriel and of "a prince of princes," Mr. Knohl contends that the stone's writings are about the death of a leader of the Jews who will be resurrected in three days.

He says further that such a suffering messiah is very different from the traditional Jewish image of the messiah as a triumphal, powerful descendant of King David.
"This should shake our basic view of Christianity," he said as he sat in his office of the Shalom Hartman Institute in Jerusalem where he is a senior fellow in addition to being the Yehezkel Kaufman Professor of Biblical Studies at Hebrew University. "Resurrection after three days becomes a motif developed before Jesus, which runs contrary to nearly all scholarship. What happens in the New Testament was adopted by Jesus and his followers based on an earlier messiah story."

Moshe Bar-Asher, president of the Israeli Academy of Hebrew Language and emeritus professor of Hebrew and Aramaic at the Hebrew University, said he spent a long time studying the text and considered it authentic, dating from no later than the first century B.C. His 25-page paper on the stone will be published in the coming months. Regarding Mr. Knohl's thesis, Mr. Bar-Asher is also respectful but cautious. "There is one problem," he said. "In crucial places of the text there is lack of text. I understand Knohl's tendency to find there keys to the pre-Christian period, but in two to three crucial lines of text there are a lot of missing words."

Moshe Idel, a professor of Jewish thought at Hebrew University, said that given the way every tiny fragment from that era yielded scores of articles and books, "Gabriel's Revelation" and Mr. Knohl's analysis deserved serious attention. "Here we have a real stone with a real text," he said. "This is truly significant." Mr. Knohl said that it was less important whether Simon was the messiah of the stone than the fact that it strongly suggested that a savior who died and rose after three days was an established concept at the time of Jesus. He notes that in the Gospels, Jesus makes numerous predictions of his suffering and New Testament scholars say such predictions must have been written in by later followers because there was no such idea present in his day.

c. Transcription of the text in English

Text (Semitic sounds in caps and\or italics)

Column A
(Lines 1-6 are unintelligible)

7. [...]the sons of Israel ...[...]...
8. [...]... [...]...
9. [...]the word of YHW[H ...]...[...]
10. [...]... I\you asked ...
11. YHWH, you ask me. Thus said the Lord of Hosts:
12. [...]... from my(?) house, Israel, and I will tell the greatness(es?) of Jerusalem.
13. [Thus] said YHWH, the Lord of Israel: Behold, all the nations are
14. ... against(?)\to(?) Jerusalem and ...,
15. [o]ne, two, three, fourty(?) prophets(?) and the returners(?),
16. [and] the Hasidin(?). My servant, David, asked from before Ephraim(?)
17. [to?] put the sign(?) I ask from you. Because He said, (namely,)
18. [Y]HWH of Hosts, the Lord of Israel: ...
19. sanctity(?)\sanctify(?) Israel! In three days you shall know, that(?)\for(?) He said,
20. (namely,) YHWH the Lord of Hosts, the Lord of Israel: The evil broke (down)
21. before justice. Ask me and I will tell you what 22this bad 21plant is,
22. lwbnsd/r/k (=? [To me? in libation?]) you are standing, the messenger\angel. He
23. ... (= will ordain you?) to Torah(?). Blessed be the Glory of YHWH the Lord, from
24. his seat. "In a little while", qyTuT (=a brawl?\ tiny?) it is, "and I will shake the
25. ... of? heaven and the earth". Here is the Glory of YHWH the Lord of
26. Hosts, the Lord of Israel. These are the chariots, seven,
27. [un]to(?) the gate(?) of Jerusalem, and the gates of Judah, and ... for the sake of
28. ... His(?) angel, Michael, and to all the others(?) ask\asked
29. .... Thus He said, YHWH the Lord of Hosts, the Lord of
30. Israel: One, two, three, four, five, six,
31. [se]ven, these(?) are(?) His(?) angel .... 'What is it', said the blossom(?)\diadem(?)
32. ...[...]... and (the?) ... (= leader?/ruler?), the second,
33. ... Jerusalem.... three, in\of the greatness(es?) of
34. [...]...[...]...
35. [...]..., who saw a man ... working(?) and [...]...
36. that he ... [...]... from(?) Jerusalem(?)
37. ... on(?) ... the exile(?) of ...,
38. the exile(?) of ..., Lord ..., and I will see

39. ...[...] Jerusalem, He will say, YHWH of
40. Hosts, ...
41. [...]... that will lift(?) ...
42. [...]... in all the
43. [...]...
44. [...]...

Column B
(Lines 45-50 are unintelligible)

51. Your people(?)\with you(?) ...[...]
52. ... the [me]ssengers(?)\[a]ngels(?)[...]...
53. on\against His/My people. And ...[...]...
54. [...]three days(?). This is (that) which(?) ...[...]He(?)
55. the Lord(?)\these(?)[...]...[...]
56. see(?) ...[...]
57. closed(?). The blood of the slaughters(?)\sacrifices(?) of Jerusalem. For He said, YHWH of Hos[ts],
58. the Lord of Israel: For He said, YHWH of Hosts, the Lord of
59. Israel: ...
60. [...]... me(?) the spirit?\wind of(?) ...
61. ...[...]...
62. in it(?) ...[...]...[...]
63. ...[...]...[...]
64. ...[...]... loved(?)/... ...[...]
65. The three saints of the world\eternity from\of ...[...]
66. [...]... peace he? said, to\in you we trust(?) ...
67. Inform him of the blood of this chariot of them(?) ...[...]
68. Many lovers He has, YHWH of Hosts, the Lord of Israel ...
69. Thus He said, (namely,) YHWH of Hosts, the Lord of Israel ...:
70. Prophets have I sent to my people, three. And I say
71. that I have seen ...[...]...
72. the place for the sake of(?) David the servant of YHWH[...]...[...]
73. the heaven and the earth. Blessed be ...[...]
74. men(?). "Showing mercy unto thousands", ... mercy [...].
75. Three shepherds went out to?/of? Israel ...[...].
76. If there is a priest, if there are sons of saints ...[...]
77. Who am I(?), I (am?) Gabri'el the ...(=angel?)... [...]
78. You(?) will save them, ...[...]...
79. from before You, the three si[gn]s(?), three ...[....]
80. In three days li[ve], I, Gabri'el ...[?],
81. the Prince of Princes, ..., narrow holes(?) ...[...]...
82. to/for ... [...]... and the ...
83. to me(?), out of three - the small one, whom(?) I took, I, Gabri'el.
84. YHWH of Hosts, the Lord of(?)[Israel ...]...[....]

85. Then you will stand ...[...]...
86. ...\
87. in(?) ... eternity(?)/... \

d. Strong Criticism

Although these findings are exciting, several caveats are in order: 1) The essential aspects of the provenance of the stone are unknown. However, Knohl states, "The authenticity of the inscription was recently checked and confirmed by Yuval Goren of Tel Aviv University." The Goren report, however, has yet to be published. 2) The stone is in bad condition, and the ink on the stone is highly faded and indecipherable in places. The Knohl Text involves some educated conjecture. 3) Knohl is known for his book, The Messiah before Jesus: The Suffering Servant of the Dead Sea Scrolls (translated in English, 2000, University of California Press), which describes the early historical context of the Jewish concept of a slain Messiah. His natural biases to the "pre-Jesus Messiah" thesis possibly could have predisposed his interpretations of the Hazon Gabriel text. 4) Knohl makes it clear that the "prince of princes" in the text is NOT identified as Jesus. However, portions of the text do appear to have a prophetic tone. 5) Scholarly analysis of the stone and its text is at the beginning stages, and there will be much debate over the Knohl Text.

On the one hand, for those who are uncomfortable with the concept of someone other than Christ acting as a suffering servant providing an intercessory offering, simply read the original atonement statute in Lev 16. It speaks of two perfect goats. One offers a sin offering (Christ) while the other offers another type of scapegoat offering. (The One like Moses) But, on the other hand, the Hazon Gabriel text surfaced on the antiquities market with only the vaguest intimation that the text was discovered in the Transjordan. This means that it has the absolute lowest level of scholarly value. Without provenance, much about location, context and use cannot be established. More fundamentally, it is impossible to know with certainty that the text is authentic. The really important discoveries, texts like the Tel Dan inscription the Black Obelisk of Shalmaneser III were discovered in situ, placing their authenticity beyond a doubt and profoundly authorizing all the provocative theses which they have inspired. Without context, the Hazon Gabriel text will never be able to sustain the argumentation it inspires. Nor, for that matter, can it inspire discussion beyond the words it bears; important questions about who produced it, for what purpose, when, where, and how, will necessarily remain speculative, and our understanding of the practices of text production in ancient Syria-Palestine will not be advanced by the discovery.

It's ink on stone, which is (as far as I know) absolutely unprecedented in ancient Syro-Palestine. We have ink on ceramic (called ostraca, found literally all over the place), ink on papyrus or leather (the Dead Sea Scrolls being a celebrated instance), we even have ink on plaster (the Deir Alla text, discovered in the Transjordan), and of course we have inscribed stone aplenty.

The Dead Sea Jesus

e. Some Other "Shaken Stones": the fragment 7Q5 and 7Q4

In 1972, the Spanish papyrologist José O'Callaghan published a controversial article, "¿Papiros neotestamentarios en la cueva 7 de Qumrán?" in which he argued that the fifth manuscript from the seventh cave of Qumran was a fragment from the Gospel of Mark (6:52-53). This produced a spate of scholarly reviews and interactions—most of which rejected O'Callaghan's identification. This rejection rested on three grounds: (1) principally, the papyrus itself was so fragmentary that *any* identification would be tenuous at best (not to mention the fact that there were several textually intrinsic problems with O'Callaghan's proposal); (2) since the Qumran community almost certainly disbanded in 68 CE—and hence the MS must be dated before that time (in fact, most likely, no later than 50 CE)—the majority of NT scholars felt that even the original draft of Mark's Gospel was not this early, obviously precluding the possibility that a *copy* of Mark could have existed before the fall of Jerusalem; and (3) the differences between the Qumran community (usually considered to be identical with the Essenes) and the nascent Christian community are so pronounced that contact between the two seemed improbable (and a *literary* contact, as O'Callaghan proposed, seemed to imply that not only was there communication between the two groups, but open and somewhat friendly communication).

O'Callaghan defended his views against virtually every assailant. But until 1982 he found few, if any, real followers. In that year Carsten Peter Thiede, a German scholar, began to publish in defense of the O'Callaghan hypothesis. In the last dozen years, in fact, he has surpassed his mentor in periodical proliferation. The book under review is, in many respects, the culmination of his efforts. *The Earliest Gospel Manuscript?*, Thiede's first book in English on the subject, has been written to appeal to a wider audience (since his earlier writings have almost completely fallen on deaf German ears). There is today both interest in and sympathy toward the O'Callaghan hypothesis—especially now that it has a fresh advocate in Thiede.[6] Indeed, at the ETS national meeting in November 1992, even Alan Johnson pleaded the case for Thiede's volume.

Why all the furor? A number of things: (1) If this identification is correct, it would be the earliest NT MS by some 50-100 years; (2) on paleographical grounds, since the *upper* limit of its date is 50 CE, this would put Mark in the 40's at the latest; (3) one consequence of such an early date for Mark would be to virtually silence advocates of Matthean priority; and (4) finally, it would suggest, perhaps, that at least some of the New Testament documents were regarded highly enough to be copied soon after publication—a view which lends itself to an early recognition of the NT as canon.[9]

The fragment "has so few words, and of such little significance (e.g., Greek *kai* = 'and') that most New Testament scholars would now appear to firmly reject this identification on the grounds that the fragment could as well be from the *Illiad* or other works of ancient Greek literature. Since 7Q5 was written in Zierstil (ornamental style), a style used from 50 B.C. to 50 A.D. (this was the dating of the noted Oxford University paleographer, Colin H. Roberts), the fragment was necessarily datable to around 40-50 A.D. (It had to be a few years after the death of Jesus, but prior to 50 A.D.) Moreover, it was clear to O'Callaghan 7Q5 could not be dated later than 68 A.D., the year the Qumran caves had been sealed by the Decima Legio Pretensis (Vespasian's Roman legion). In that year, Vespasian, marching toward Jerusalem, had arrived at the Dead Sea and ordered his troops to fan out and massacre the small Jewish monastic communities of the area.

"For they did not understand about the loaves, but their hearts were hardened. And when they had crossed over, they came to land at Gennesaret and moored to the shore." (Mark 6:52-53)

The ongoing debate about the contents of Cave 7 and the Markan papyrus 7Q5 tends to overshadow the fact that there is a papyrus scroll fragment from this cave (7Q4) which was identified by Father O'Callaghan as 1 Timothy 3:16-4:3. This claim has proved resilient in the face of skepticism muted as that has been. Indeed, even scholars who remained skeptical as to the identification of 7Q5 were convinced by the identification of 7Q4 as verses from 1 Timothy.

Putting all this in perspective, we conclude this review by addressing two concerns: evidence and attitudes. First, what is the hard evidence on which O'Callaghan's identification is based? A scrap of papyrus smaller than a man's thumb with only one unambiguous word—kai. Only six other letters are undisputed: tw (line 2), t (line 3, immediately after the kai), nh (line 4), h (line 5). To build a case on such slender evidence would seem almost impossible even if all other conditions were favorable to it. But to identify this as Mark 6:52-53 requires (1) two significant textual emendations (*tau* for *delta* in a manner which is unparalleled; and the dropping of ejpiV thVn gh'n even though no other MSS omit this phrase); and (2) unlikely reconstructions of several other letters. Add to this that the MS is from a *Qumran* cave and that it is to be dated no later than 50 CE and the case *against* the Marcan proposal seems overwhelming. If it were not for the fact that Jos O'Callaghan is a reputable papyrologist and that C. P. Thiede is a German scholar, one has to wonder whether this hypothesis would ever have gotten more than an amused glance from the scholarly community.

Second, regarding attitude, I find it disturbing that many conservatives have been so uncritically eager to accept the O'Callaghan hypothesis. 7Q5 does not, as one conservative put it, mean "that seven tons of German scholarship may now be consigned to the flames." On the other hand, I find it equally disturbing that many liberal scholars have uncritically rejected O'Callaghan's proposal without even examining the evidence. Higher criticism must of course have a say in this discussion; but it must not *preclude* discussion. Both attitudes, in their most extreme forms, betray an arrogance, an unwillingness to learn, a fear of truth while clinging to tradition, a fortress mentality—none of which is in the spirit of genuine biblical scholarship.

The Scrolls From the Dead Sea

Survey of the Caves

While many of the Dead Sea Scrolls are small fragments of Biblical, apocryphal, or sectarian manuscripts, some of the scrolls have come to be well known and influential to Second Temple Judaism. The caves surrounding Qumran are numbered based upon the order of their discovery and their production of scrolls and scroll fragments. Thus, caves 7-9 and 4 are very close to the settlement at Qumran, while caves 1, 3, and 11 are farther away. Likewise, there are hundreds of other caves surrounding Qumran discovered both before and after the 11 scroll caves that did not produce scrolls and are therefore not numbered as scroll caves. Below is a summary of each of the Qumran Caves:

Scrolls from Cave 1

1QIsaa (a copy of the book of "Isaiah")
1QIsab (a second copy of the book of "Isaiah")
1QS ("Serekh ha-Yahad" or "Community Rule") cf. 4QSa-j = 4Q255-64, 5Q11
1QpHab ("Pesher on Habakkuk")
1QM ("Milhamah" or "War Scroll") cf. 4Q491, 4Q493; 11Q14?
1QHa ("Hodayot" or "Thanksgiving Hymns")
1QapGen ar ("Genesis Apocryphon" in Aramaic)
CTLevi ar ("Cairo Geniza Testament of Levi" in Aramaic)
1QGen ("Genesis") = 1Q1
1QExod ("Exodus") = 1Q2
1QpaleoLev ("Leviticus" written in palaeo-Hebrew script) = 1Q3
1QDeuta ("Deuteronomy") = 1Q4
1QDeutb ("Deuteronomy") = 1Q5
1QJudg ("Judges") = 1Q6
1QSam ("Samuel") = 1Q7
1QIsab (fragments from the 1QIsab scroll) = 1Q8
1QEzek ("Ezekiel") = 1Q9
1QPsa ("Psalms") = 1Q10
1QPsb ("Psalms") = 1Q11
1QPsc ("Psalms") = 1Q12
1QPhyl (58 fragments from a "Phylactery") = 1Q13
1QpMic ("Pesher on Micah") = 1Q14
1QpZeph ("Pesher on Zephaniah") = 1Q15
1QpPs ("Pesher on Psalms") = 1Q16
1QJuba ("Jubilees") = 1Q17
1QJubb ("Jubilees") = 1Q18
1QNoah ("Book of Noah") = 1Q19
1QapGen ar (fragments from the "Genesis Apocryphon" in Aramaic) = 1Q20
1QTLevi ar ("Testament of Levi" in Aramaic) = 1Q21
1QDM ("Dibrê Moshe" or "Words of Moses") = 1Q22
1QEnGiantsa ar ("Book of Giants" from "Enoch" text in Aramaic) = 1Q23
1QEnGiantsb ar ("Book of Giants" from "Enoch" text in Aramaic) = 1Q24

1Q25 ("Apocryphal Prophecy")
1Q26 ("Instruction")
1QMyst ("Mysteries") = 1Q27
1Q28 (fragment of the title of "1QS" or "Community Rule")
1QSa ("Rule of the Congregation") = 1Q28a
1QSb ("Rule of the Blessing" or "Rule of the Benedictions") = 1Q28b
1Q29 ("Liturgy of the Three Tongues of Fire")
1Q30 ("Liturgical Text")
1Q31 ("Liturgical Text")
1QNJ ar ("New Jerusalem" text in Aramaic) = 1Q32 cf. 11Q18 ar
1Q33 (fragment of 1QM or "War Scroll")
1QLitPr ("Liturgical Prayers" or "Festival Prayers") = 1Q34
1QHb ("Hodayot" or "Thanksgiving Hymns") = 1Q35
1Q36-40 ("Hymnic Composition")
1Q41-70 (Unclassified Fragments)
1QDana ("Daniel") = 1Q71
1QDanb ("Daniel") = 1Q72

Scrolls from Cave 2

2QGen ("Genesis") = 2Q1
2QExoda ("Exodus") = 2Q2
2QExodb ("Exodus") = 2Q3
2QExodc ("Exodus") = 2Q4
2QpaleoLev (section of "Leviticus" 11:22-29 written in palaeo-Hebrew script) = 2Q5
2QNuma ("Numbers") = 2Q6
2QNumb ("Numbers") = 2Q7
2QNumc ("Numbers") = 2Q8
2QNumd ("Numbers") = 2Q9
2QDeuta ("Deuteronomy") = 2Q10
2QDeutb ("Deuteronomy") = 2Q11
2QDeutc ("Deuteronomy" 10:8-12) = 2Q12
2QJer ("Jeremiah") = 2Q13
2QPs ("Psalms") = 2Q14
2QJob ("Job" 33:28-30) = 2Q15
2QRutha ("Ruth") = 2Q16
2QRuthb ("Ruth") = 2Q17
2QSir ("Sirach" or "Wisdom of Jesus ben Sira" or "Ecclesiasticus") = 2Q18
2QJuba ("Jubilees") = 2Q19
2QJubb ("Jubilees") = 2Q20
2QapMoses ("Apocryphon of Moses") = 2Q21
2QapDavid? ("Apocryphon of David?") = 2Q22
2QapProph ("Apocryphal Prophecy") = 2Q23
2QNJ ar ("New Jerusalem" text in Aramaic) = 2Q24 cf. 1Q32 ar, 11Q18 ar
2Q25 ("Juridical Text")

2QEnGiants ar ("Book of Giants" from "Enoch" in Aramaic) = 2Q26 cf. 6Q8
2Q27-33 (unidentified texts)

Scrolls from Cave 3

3QEzek ("Ezekiel" 16:31-33) = 3Q1
3QPs ("Psalms" 2:6-7) = 3Q2
3QLam ("Lamentations") = 3Q3
3QpIsa ("Pesher on Isaiah") = 3Q4
3QJub ("Jubilees") = 3Q5
3QHymn (an unidentified hymn) = 3Q6
3QTJudah? ("Testament of Judah"?) = 3Q7 cf. 4Q484, 4Q538
3Q8 (fragment of an unidentified text)
3Q9 (possible unidentified sectarian text)
3Q10-11 (unclassified fragments)
3Q12-13 (unclassified Aramaic fragments)
3Q14 (unclassified fragments)
3QCopper Scroll ("The Copper Scroll") = 3Q15

Scrolls from Cave 4

4QGen-Exoda ("Genesis and Exodus") = 4Q1
4QGenb ("Genesis") = 4Q2
4QGenc ("Genesis") = 4Q3
4QGend ("Genesis" 1:18-27) = 4Q4
4QGene ("Genesis") = 4Q5
4QGenf ("Genesis" 48:1-11) = 4Q6
4QGeng ("Genesis") = 4Q7
4QGenh1 ("Genesis" 1:8-10) = 4Q8
4QGenh2 ("Genesis" 2:17-18) = 4Q8a
4QGenh-para (a paraphrase of "Genesis" 12:4-5) = 4Q8b
4QGenh-title (the title of a "Genesis" manuscript) = 4Q8c
4QGenj ("Genesis") = 4Q9
4QGenk ("Genesis") = 4Q10
4QpaleoGen-Exodl ("Genesis and Exodus" written in palaeo-Hebrew script) = 4Q11
4QpaleoGenm ("Genesis" written in palaeo-Hebrew script) = 4Q12
4QExodb ("Exodus") = 4Q13
4QExodc ("Exodus") = 4Q14
4QExodd ("Exodus") = 4Q15
4QExode ("Exodus" 13:3-5) = 4Q16
4QExod-Levf ("Exodus and Leviticus") = 4Q17
4QExodg ("Exodus" 14:21-27) = 4Q18
4QExodh ("Exodus" 6:3-6) = 4Q19
4QExodj ("Exodus") = 4Q20

4QExodk ("Exodus" 36:9-10) = 4Q21
4QpaleoExodm ("Exodus" written in palaeo-Hebrew script) = 4Q22
4QLev-Numa ("Leviticus and Numbers") = 4Q23
4QLevb ("Leviticus") = 4Q24
4QLevc ("Leviticus") = 4Q25
4QLevd ("Leviticus") = 4Q26
4QLeve ("Leviticus") = 4Q26a
4QLevg ("Leviticus") = 4Q26b
4QNumb ("Numbers") = 4Q27
4QCantb ("Pesher on Canticles or "Pesher on the Song of Songs) = 4Q107
4QCantc ("Pesher on Canticles or "Pesher on the Song of Songs) = 4Q108
4Q112 ("Daniel")
4Q123 ("Rewritten Joshua")
4Q127 ("Rewritten Exodus")
4Q128-148 (various tefillin)
4Q156 ("Targum of Leviticus")
4Q157 ("Targum of Job") = 4QtgJob
4QRPa ("Rewritten Pentateuch") = 4Q158
4Q161-164 ("Pesher on Isaiah")
4Q166-167 ("Pesher on Hosea")
4Q169 ("Pesher on Nahum")
4Q174 ("Florilegium" or "Midrash on the Last Days")
4Q175 ("Messianic Anthology" or "Testimonia")
4Q179 ("Lamentations") cf. 4Q501
4Q196-200 ("Tobit")
4Q213-214 ("Aramaic Levi")
4Q215 ("Testament of Naphtali")
4QCanta ("Pesher on Canticles or "Pesher on the Song of Songs") = 4Q240
4Q252 ("Pesher on Genesis")
4Q265-273 ("CD or "Damascus Document") cf. 4QDa/g = 4Q266/272, 4QDa/e = 4Q266/270, 5Q12, 6Q15, 4Q265-73
4Q285 ("Rule of War") cf. 11Q14
4QRPb ("Rewritten Pentateuch") = 4Q364
4QRPc ("Rewritten Pentateuch") = 4Q365
4QRPc ("Rewritten Pentateuch") = 4Q365a (=4QTemple?)
4QRPd ("Rewritten Pentateuch") = 4Q366
4QRPe ("Rewritten Pentateuch") = 4Q367
4Q434 ("Barkhi Napshi - Apocryphal Psalms") (15 fragments likely hymns of thanksgiving, praising God for his power and expressing thanks)
4QMMT ("Miqsat Ma'ase Ha-Torah" or "MMT" or "Some Precepts of the Law" or the "Halakhic Letter") cf. 4Q394-399
4Q400-407 ("Songs of Sabbath Sacrifice" or the "Angelic Liturgy") cf. 11Q5-6
4Q448 ("Hymn to King Jonathan")
4Q521 ("Messianic Apocalypse")
4Q539 ("Testament of Joseph")
4Q554-5 ("New Jerusalem") cf. 1Q32, 2Q24, 5Q15, 11Q18

Scrolls from Cave 5

5QDeut ("Deuteronomy") = 5Q1
5QKgs ("1 Kings") = 5Q2
5QIsa ("Isaiah") = 5Q3
5QAmos ("Amos") = 5Q4
5QPs ("Psalms") = 5Q5
5QLama ("Lamentations") = 5Q6
5QLamb ("Lamentations") = 5Q7
5QPhyl (3 fragments from a "Phylactery") = 5Q8
5QapJosh ("Apocryphon of Joshua") = 5Q9
5Q10 Apocryphon of Malachi
5Q11 Rule of the Community
5Q12 Damascus Document
5Q13 Rule
5Q14 Curses
5Q15 New Jerusalem
5Q16-25 unclassified
5QX1 Leather fragment

Scrolls from Cave 6

6QpaleoGen (section of "Genesis" 6:13-21 written in palaeo-Hebrew script) = 6Q1
6QpaleoLev (section of "Leviticus" 8:12-13 written in palaeo-Hebrew script) = 6Q2
6Q3 Deuteronomy
6Q4 Kings
6QCant ("Canticles" or "Song of Songs") = 6Q6
6Q7 Daniel
6QpapEnGiants ("Book of Giants" from "Enoch") = 6Q8
6Qpap apSam-Kgs ("Apocryphon on Samuel-Kings") = 6Q9
6QpapProph (an unidentified prophetic fragment) = 6Q10
6Q11 ("Allegory of the Vine")
6QapocProph (an apocryphal prophecy) = 6Q12
6QPriestProph ("Priestly Prophecy") = 6Q13
6QD ("Damascus Document") = 6Q15
6QpapBened ("Benediction") = 6Q16
6Q17 Calendrical Document
6Q18 Hymn
6Q19 Genesis
6Q20 Deuteronomy
6Q21 Prophetic text?
6Q22-6QX2 Unclassified

Scrolls from Cave 7

7QLXXExod (a section of "Exodus" from the Septuagint) = 7Q1
7QLXXEpJer ("Letter of Jeremiah" = Baruch 6) = 7Q2
7Q3 Biblical Text?
7QpapEn gr ("Enoch") = 7Q4, 8, 11-14
7Q5 Biblical Text
7Q6, 7, 9, 10 unclassified
7Q15-18 unclassified
7Q19 imprint

Scrolls from Cave 8

8QGen ("Genesis") = 8Q1
8QPs ("Psalms") = 8Q2
8QPhyl (fragments from a "Phylactery") = 8Q3
8QMez (portion of "Deuteronomy" 10:12-11:21 from a Mezuzah) = 8Q4
8QHymn (a previously unidentified hymn) = 8Q5
8QX1 Tabs
8QX2-3 Thongs

Scrolls from Cave 9

9Qpap (unidentified fragment)

Scrolls from Cave 10

10Q1 ostracon

Scrolls from Cave 11

11QpaleoLeva ("Leviticus" written in palaeo-Hebrew script) = 11Q1
11QpaleoLevb ("Leviticus" written in palaeo-Hebrew script) = 11Q2
11QDeut ("Deuteronomy") = 11Q3
11QEz ("Ezekiel") = 11Q4
11QPsa ("Psalms") = 11Q5
11QPsb ("Psalms") = 11Q6
11QPsc ("Psalms") = 11Q7
11QPsd ("Psalms") = 11Q8
11QPse ("Psalms") = 11Q9
11QtgJob ("Targum of Job") = 11Q10
11QapocrPs ("Apocryphal Psalms") = 11Q11
11QJub ("Jubilees") = 11Q12
11QMelch ("Heavenly Prince Melchizedek") = 11Q13
11QSM ("Sefer Ha-Milhamah" or "Book Of War") = 11Q14. cf. 1QM?
11QHymnsa = 11Q15

11QHymnsb = 11Q16
11QShirShabb ("Songs of the Sabbath Sacrifice") = 11Q17
11QNJ ar ("New Jerusalem" text in Aramaic) = 11Q18 cf. 1Q32, 2Q24
11QTa ("Temple Scroll") = 11Q19
11QTb ("Temple Scroll") = 11Q20
11Q21 Hebrew text
11Q22-28 unclassified
11Q29 Serekh ha-Yahad related
11Q30 unclassified
11Q31 unclassified
XQ1-4 Phylacteries
XQ5 & 6 fragments and offering

Rewritten Bible Texts

Fernando Klein

Genesis Apocryphon (Tales of the Patriarchs (1QapGen=1Q20)

The "Tales of the Patriarchs," which deals with the descendants of Adam, is sometimes referred to as the "Genesis Apocryphon." Originally, the *Genesis Apocryphon* was referred to as the fourth scroll because it was the fourth scroll out of seven to be found in the Qumran Cave. This "Dead Sea Scroll" was originally thought to have come from the apocryphal book of Enoch because a small portion of the scroll that had been unraveled, mentioned Enoch's name. However, when another section was unraveled, scholars were lead to believe that this scroll came from the apocryphal book of Lamech, a name that was already known to scholars because of the book *Jubilees*. The reason why they thought this scroll came from the book of Lamech is because the speaker spoke about Bitenosh, Lamech's wife, in first person (Yadin, pg. 144). Yet once again, this belief was wrong. When the scroll was finally fully unraveled, it had references to Noah, Abraham and Lot, making the scrolls relation to the *Jubilees* more apparent. In many senses one can refer to this text as a "Little Genesis" because its literary dependence on Genesis is similar to that of the *Jubilees*.

When Avigad and Yadin published the fourth scroll, they realized that they could no longer call it the "Book of Lamech" and so they decided on the title *Genesis Apocryphon* to avoid any further commitments to the character of the writing contained in the scroll. Avigad stated that "these stories are based on the biblical narratives but they also deal with other subjects and details previously unknown". An example of this is in of the story of Joseph and Sarah and their time in Egypt. In column 20 of the Apocryphon, the time duration of Sarah living with the Pharaoh is given as two years unlike the bible when a time frame was not given (Yadin, pg. 144). Also, the purity of Sarah which was in question in the bible is maintained in the *Genesis Apocryphon*. Sarah could not have relations with the Pharaoh because the evil spirit that Joseph prayed for to God made all the men in Egypt impotent. Thus, the purity of Adam descendants were maintained.

However, not all scholars agreed with this title Apocryphon because it evokes its counterpart, a canonical book, and consequently introduces not the Qumran literature a slight anachronism. Yet, even with some criticism, the fourth scroll is known as the Genesis Apocryphon.

Now that the debate of naming this scroll was over, the task of placing this text into a genre began. It has already been stated that this text relies heavily on the canonical Genesis, however, this scroll contains additional details that were obviously derived from some non-biblical sources. For instance, Gn 5.28-29 in the bible is the starting point for the extended narrative for columns 2-5 which are embellishments of the birth of Noah. Columns 6-17 deals with Noah, the flood, and the division of the earth between his sons. These lines can once again be compared to the Jubilees chapters 4-9. In column 18 and 19 where the narrative has shifted to Abraham. This text is the expanded version corresponds with of Gn 11.27-14. There is an obvious similarity between the working of this section to the Jubilees. In column 22, a less direct Text can be seen, with the author reverting to the free reworking of the Genesis story which has been seen in previous columns, yet not returning to the first person except for conversation itself.

Frag.1 Col. 1

You should let your anger and tear out (?)... and who is the man who.. the fury of your anger..... and those who have been destroyed and killed, bereft and... and now I have stopped the prisoners.... the Great Holy One....all that he...

Frag.1 Col 2:

day of... all...land of... and the evil for...

Frag. 2

...and they were hit from behind...in front of the lord

Col. 1

.... and with the sowing....not even the mystery of evil which....the mystery which

Col. 2.

I thought, in my heart, that the conception was the work of the Watchers the pregnancy of the Holy Ones and that it belonged to the Giants... and my heart was upset by this... I, Lamech, turned to my wife Bitenosh and said... Swear to me by the Most High, Great Lord, King of the Universe...the sons of heavens, that you will truthfully tell me everything, if... You will tell me without lies... Then Bitenosh, my wife spoke harshly and she cried... and said: Oh my brother and lord! Remember my pleasure... the time of love, gasping for breath. I will tell you everything truthfully... and then my heart began to ache... When Bitenosh realized my mood had changed...Then she withheld her anger and said to me: O my lord and brother! Remember my pleasure. I swear to you by the Great Holy One, the King of the heavens... That this seed, pregnancy, and planting of fruit comes from you and not a stranger, Watcher, or son of the heaven... Why is your expression changed and your spirit saddened... I speak honestly to you... Then I, Lamech, went to my father, Methuselah, and told him everything so that he would know the truth because he is well liked... and he is in well with the Holy Ones and they share everything with him. Methuselah went to Enoch to find the truth... he will. And he went to Parvaim, where Enoch lived... He said to Enoch: O my father and lord, to whom I... I tell you! Do not be angry because I came here to you... fear before you...

Fernando Klein

Col 3

For in the days of Jared, my father...

Col 5

Enoch...not from the sons of heaven, but from Lamech your son... I now tell you... and I reveal to you... Go tell your son Lamech... When Methuselah heard this... And with his son Lamech, he spoke... Now when I, Lamech, heard these things... Which he got out of me

Col 6

I abstained from injustice and in the womb of my mother who conceived me I searched for truth. When I emerged from my mother's womb, I lived all my days in truth and walked in the path of eternal truth. And the Holy One was with me... on my pathways truth sped to warn me off the... of lie which led to darkness.. I braced my loins with the vision of truth and wisdom... paths of violence. *vacat* Then, I Noah became a man that clung to truth and seized... I took Amzara, his daughter as my wife. She conceived and bore me three sons and daughters. I Then took wives from my brother's family for my sons, and I gave my daughters to my nephews according to the law of the eternal precept which Most High ordained to the sons of man. *vacat* And in my days, when according to my reckoning... ten jubilees had been completed, the time came for my sons to take wives for themselves... heaven, I saw in a vision and was explained and made known the actions of the sons of heaven and... the heavens. Then I hid this mystery in my heart and explained it to no one. *vacat*... to me and a great and... and in a message of the Holy One... and he spoke to me in a vision and he stood before me... and the message of the Great Holy One called out to me: "To you they say, O' Noah,..." and I reckoned the whole conduct of the sons of the earth. I knew and explained everything... two weeks. then the blood which the Giants had spilled... I was at ease and waited until... the holy ones with the daughters of man... The I Noah, found grace, greatness and for my entire life I have behaved righteously...I, Noah, a man...

Col. 7

God told Noah that he would rule over the earth and the seas and all they encompass. Noah was overjoyed at the idea.

Col. 10

The arc rested upon the mountain of Ararat (Hurarat). Noah atoned for the land and burned incense on the alter.

Col. 11

God makes a covenant with Noah telling him he could no longer eat blood of any kind.

Col. 12

I placed my bow in the cloud and it became a sign for me in the cloud... the earth... it was revealed to me in the mountains... a vineyard in the mountains of Ararat... After the flood Noah and his sons descended from the mountain. They saw the widespread devastation of the earth. After the flood Noah's children began to have his grandchildren-Sons ands daughters. They then planted the soil and put a vineyard on Mount Lubar that produced wine four years later: On the first day of the fifth year, there was a feast at which the first wine was drank. Noah gathered his family together and they went to the alter and thanked god for saving them from the destruction of the flood.

Col 13

...They were cutting gold, silver, stones, and clay and taking some for themselves. I saw the gold and silver... iron, and they cut down all the trees and took some. I saw the sun, moon, and stars cutting and taking some for themselves... I turned to see the olive tree and behold, it was rising up and for many hours... many leaves... appeared in them. I watched the olive tree and the abundance of its leaves... they tied to it. I was greatly amazed by the tree and its leaves... the four winds of heaven were strongly blowing and they were breaking off and smashing the branches of the olive tree. The westerly wind hit first, knocking off its fruit and leaves and scattering them everywhere. Then...

Col 14

...Listen and hear! You are the great cedar... standing in front of you in a dream on the mountain tops... truth. The willow that springs from it and rises high (these are) three sons... And the one that you did see, the first willow got attached to the stump of the cedar... and the wood from it... will never separate from you. And among it posterity... will be called... will grow a wonderful plant... will stand forever. And what you saw, the willow caught the stump... the last willow... part of their branch entered the branch of the first tree, two sons... And what you saw, that part of their branch entered the branch of the first tree... I explained to him the mystery...

Col 15

...And that you saw all of them... They will go around, the majority of them will be evil. and what you saw, that a man came from the south, with a sickle in his hand, and bringing fire with him... who will come from the south of the land... And they will put wickedness on the fire, a;;... And he should come between... Four angels... between all nations. And they will all worship and be dumbfounded... I will honestly explain to you. And I, Noah woke up from my sleep and the Sun.

Col. 16

Noah divided the land among his decedents.... all the land of the north as far as... this boundary, the waters of the Mediterranean.... the Tina River.

Col. 17

Noah further divided the land West, to Asshur, as far as the Tigris. He gave Aram land as far as the source of.... this Mountain of the Bull, and he crossed it westward as far as.... where the three parts met.... For Arpachshad... He gave Gomer a part in the northeast t the Tina River.... To Magog...

Col. 19

I, Abraham built and alter (at Bethel) and called to god, praising him. I then went to the Holy mountain and to Hebron where he lived for two years. Because there was famine in the land my family and I traveled to Egypt where grain was plentiful. I went across the branches of the Nile to enter Egypt, the land of the sons of Ham. I had a dream about a cedar tree and a date-palm tree. When people came to cut down the cedar tree, the date-palm tree objected, saying that they were grown from a single root. The cedar tree was spared. I became fearful of the dream and told it to my wife. I explained it as it pertained to us telling Sarah that the men will come for her and try to kill me. I warned Sarah that she must tell everyone that I am her brother so that my life can be spared. She became scared and did not want to go to Zoan for fear of being seen. Five years later, councilors of the Egyptians court and advisors of the Pharaoh of Zoan came, having heard the words of my wife. They brought gifts and requested knowledge from me. I read to them from the Book of the words of Enoch.

Col. 20

The men return to the Pharaoh and describe Sarah's features: beautiful face, supple hair, lovely eyes, pleasant nose, radiant face. He continued on describing her shapely breasts, perfect hands, and everything down to her long and delicate fingers. the men compared her to and rated her far higher than virgins and birds, and all other women alike. Hearing this, and then seeing Sarah, the pharaoh wanted her and took her for his wife. Sarah saved me by telling the pharaoh that I was her brother and that night I and my nephew Lot cried together I prayed to Lord for justice. I wanted the Lord to raise up against the pharaoh and protect Sarah. God listened and sent an evil spirit to the entire household that prevented the pharaoh from having sexual relations with Sarah for the two years that they were together. At the end of the two years, the plagues and afflictions were so great that magicians and healers were sent for. They were, of course, ineffective, and they all soon left. Hyrcanos went to me pleading for help against the plague because I had been seen in a dream. I agreed to help only when my wife Sarah is returned to me. The pharaoh heard this and confronted me, himself asking why I lied saying that Sarah was my sister. He agreed to give Sarah back and I exorcised the evil spirit from the house of the pharaoh. The pharaoh swore to me that he had not touched Sarah while they were together and gave her gifts of gold, silver, linen, and purple-dyed clothing. Sarah and I were then led out of Egypt. I, Sarah, Lot, and his wife took our flocks and the gold and silver I had received and traveled together.

Col. 21

I went to all my old campsites until I reached Bethel, the place where I once built an alter, and then I built another one and offered up burnt offerings and a cereal offerings to the God Of Most High, and invoked the name of the Lord of the Universe there. I praised God's name and blessed god and gave thanks to Him there for all the flocks and goods and wealth which he has given me, for the good he has done for me, and because He had returned me to this land safely.

After this day, Lot left me on the account of our shepherd's behavior. He went to live in the Valley of Jordan taking all his flocks with him. And I also added greatly to what he had. he pastured his flock and kept moving until he reached Sodom and bought a house there, while I still lived in the mountain of Bethel. It bothered me that Lot and I had separated.

God came to me in a dream and said to me: Go up to Ramat Hazor which is north of Bethel, the place you are living now, and look to the east, west, south and to the north. Look at the land which I am giving you and your descendants forever. The next morning I went up to Ramat Hazor and looked at the land from that height, from the river of Egypt up to Lebanon and Senir, and from the Great Sea up to Hauran, and all the land of Gebel to Qadesh, and all the Great Desert, as far as the Euphrates and he said to me: I shall give all this land to your descendants; and they will inherit it forever. I will multiply your descendants like the dust of the earth that none can count. Your descendants will be numberless. Arise, walk about, go "see how long and how wide it is, for I will give it to you and to your descendants after you, forever.

Then I, Abraham, went out traveling in a circuit to survey the land. I began the circuit at the Gihion River, I went along the Mediterranean Sea until I reached the Mountain of the Bull. I circled from the coast of this great river saltwater sea, skirting the Mount of the Bull, and continued eastward through the breadth of the and until I came to the Euphrates river. I traveled along the Euphrates until I reached the red sea in the east, whence I followed the coast of the Red Sea until I came to the branch of the Reed Sea, jutting out from the Red Sea. From there I completed the circuit, moving southward to arrive at Gihon River. Then I returned home safely and found all is well with my men. Then I went and settled next to the oaks of Mamre, which is northeast of Hebron. There I built an alter and offered up burnt offering and a cereal offering to the God Most High. I ate and drank there, I and all the men of my household, and invited Mamre, Arnem, and Eshkol, three Amorite brothers and my friends. They ate and drank together with me. Prior to those days Chedorlaomer, the king of Elam, Amraphel, the king of Babylon, Arioch, the king of Cappadocia, and Tidal, the king of Goiim, which lies between the two rivers had come. They had waged war on Bera, the king of Sodom, Birsha, the king of Gomorrah, Shinab, the king of Admah, Shemiabad, the king of Zeboiim, and the king of Bela. All these formed an alliance to do battle in the Valley of Siddim. Now the king of Elam, and the kings with him proved to be stronger than the king of Sodom and imposed tribute upon them. Over twelve years they continued paying their tribute to the king of Elam, but in the thirteenth they rebelled against him. Thus the fourteenth year the king of Elam sallied forth with all his allies, and they ascended by the way of the desert. They smote and plundered beginning from the Euphrates. They kept on smiting-smiting the Rephaim who were in the Asteroth- Kernaim, the Zumzammin who were Amman, the Emim who were in Shaveh- hakerioth, and the Horites who were in the mountain of Gebal-until they reached El- Paran, in the desert. They returned...in Hazazon-tamar. The king of Sodom went out to meet him, together with the king of Gomorrah, Admah, Zeboiim and the king of Bela. They engaged in battle in the valley of Siddim against Chedorlaomer, and allies that were with him. The king of Sodom was defeated and put to flight while the king of Gomorrah fell into the pits... The king Elam plundered all the property of Sodom and of Gomorrah and they captured Lot.

A Genesis Florilegium (4 Q252)

This text is one of the most fascinating in the corpus. It consists of some six columns as we have reconstructed it and skims over the main Genesis narrative, alighting only on points and issues it wishes for some reason to clarify or re-present. These include the flood, Ham's son Canaan's punishment, the early days of Abram/Abraham, Sodom and Gommorah, and Reuben's offence against his father. It ends, perhaps most importantly, with Jacob's blessing of his children. This last, more of an interpretation (pesher) than a rewrite, incorporates some of the most telling Messianic pronouncements of any Qumran text in this or any other volume.

In the Genesis Florilegium, there is a collateral interest in sexual matters reflecting the condemnation of 'fornication' which one finds in other Qumran documents like that in the 'three nets of Belial' section of the Damascus Document. This is a main concern of James' instructions to overseas communities in Acts, as it is in the letter attributed to his name. This concern is not only prominent in both the Ham/ Canaan and Sodom/Gomorrah episodes before us, but also the stories which follow these about blotting out Amalek's name 'from under Heaven' and Reuben's disqualification from his rightful legacy owing to his sexual relations with his father's concubine Bilha. This latter was seemingly as jarring to ancient ears as it is to modern. The text ends in Column 6, a little anticlimactically, with portions from Gen. 49:20-21 about blessings on Asher and Napthali, of which little is intelligible.

Text

Column 1

(1) in the 480th [year] of Noah's life their (Wicked humanity) end came for Noah. And *God* (2) [sa]id, 'My Spirit shall not dwell among men forever,' and so their days were fixed (at) one hundred and twenty (3) [yea]rs, until the time of the waters of the flood. Now the waters of the flood were on the earth beginning with the six hundredth year (4) of Noah's life. In the second month, on Sunday the 17th, on that very day (5) all the fountains of the great deep burst open, and the windows of Heaven were opened. So there was rain on (6) the earth for forty days and forty nights, until the 26th of the third (7) month, Thursday. The wa[te]rs rose upon the [ea]rth for one hundred and fifty days, (8) until the 14th of the seventh month, Tuesday. And at the end of one hundred (9) and fifty days the waters abated, for two days, Wednesday and Thursday, and on (10) Friday the ark came to rest on the Ararat Range-the 17th of the seventh month. (11) Now the waters [con]tinued to diminish until the [ten]th month. On the first of that month, Wednesday, (12) the peaks of the mountains bec[ame visible. Forty days from the ti[me] when the mou[ntain] peaks became visible, (13) Noah [ope]ned the window of the ark. On Sunday, that is, the 10th of (14) the [eleve]nth month, [No]ah sent forth the dove to see whether the waters had abated, but (15) it did not find any place to alight and so it returned to him [in t]he ark. He then waited seven [mor]e days (16) and once more sent it forth, and it returned to him with a cut olive branch in its bill. [This was on the (17) 2]4th of the eleventh month, on Sunda[y. Therefore Noah knew that the waters had abated] (18) on the earth. At the end of seven mo[re days Noah sent the dove out, but (19) it did not] return again. This was the fir[st day of the twelfth] month, [a (20) Sunday.] At the end of thirt[y-one days from the time he had sent it forth], when it did not (21) return anymore, the wa[ters] had dried up [on the earth.] Then Noah removed the hatch of the ark (22) and looked around, and indeed [the waters had disappeared from the face of the earth], on the first day of the first month,

Column 2

(1) in the six hundred and first year of Noah's life. And on the 17th of the second month, (2) the earth was completely dry. On Sunday, on that day Noah went forth from the ark, thus completing a full (3) year of three hundred and sixty four days. On Sunday, in the seventh (4) <one and six. > Noah (went forth) from the ark at the appointed time, one full (5) year. < > 'Then Noah awoke from his wine and knew what his youngest son (6) had done to him, and he said, "Cursed be Canaan; he shall be his brothers' meanest slave."' He did not (7) curse Ham, but on the contrary, his son, because *God* had already blessed Noah's sons: 'And in the tents of Shem they will dwell.' (8) He gave the land to Abraham His friend. < > Terah

was one hundred and f[o]rty years old when he left (9) Ur of the Chaldees and came to Haran. And Ab[ram was seventy, and Abram lived in (10) Haran for five years, and after [Abram] left [for] the land of Canaan, (Terah lived) sixt[y-five years...] (11) the heifer and the ram and the sheg[oat...] Abram to *God*... (12) the fire when he crossed over... (13) Abr[am] to go out [to the land of] Canaan...

Column 3

(1) as it is written .... twelve (2) me[n... Gomor]rah and also (3) this city... Righteous (4) I will not destroy... only they shall exterminate. (5) And if there are not found there [ten Righteous Men, I will destroy the city and everyone] found in it, along with its booty (6) and its little children. And the remnant... forever. And Abraham (7) stretched out his hand... (8) And he said to him, 'No[w...'] (12) 'And El Shaddai will bless you and make you fruitful and multiply you. You shall become a congregation of peoples. And he will give to you (13) the blessing once given to [Abraham] your father'...

Column 4

(1) '[...and] Timna was the concubine of Eliphaz the son of Esau, and she bore him Amalek.' It was he whom Saul exterminat[ed] (2) as He said to Moses, 'In the future you will erase the memory of Amalek (3) from under Heaven.' < > The blessings of Jacob: 'Reuben, you are my first born, (4) the first portion of my strength, preeminent in stature and preeminent in power, unstable as water-(but) you shall not be preeminent. You mounted (5) your father's marriage couch, thereby defiling it because he lay on it.' < > Interpreted, this means that he reproved him, because (6) he (Reuben) slept with Bilhah his (father's) concubine. When it says 'You are my first born,' it means... Reuben was (7) the first in theory...

Column 5

(1) '(the) Government shall [not] pass from the tribe of Judah.' During Israel's dominion, (2) a Davidic descendant on the throne shall [not c]ease. For 'the Staff' is the Covenant of the Kingdom. (3) [The leaders of Israel, they are 'the Feet' (referred to in Genesis 49:11), until the Messiah of Righteousness, the Branch of (4) David comes, because to him and his seed was given the Covenant of the Kingdom of His people in perpetuity, because (5) he kept... the Torah with the men of the Community, because (6)... refers to the Congregation of the men of (7)... He gave

Column 6

(1) 'he shall yield royal dignities. Naphtali is a doe let loose, who gives (2) beautiful words.'...

Reworked Pentateuch (4QPPa=4Q158)

4Q158 is also known as 4QPP[a], and The Reworked Pentateuch[a]. 4Q158 is grouped with 4Q364-7, and together are called the Reworked Pentateuch. 4Q158 contains portions of Genesis, Exodus, and Deuteronomy. Most lines are exactly as they appear in the Bible and some are extrabiblical. In general, 4Q158 parallels quotes from the Pentateuch with minor additions. Some scholars wonder if it could be an atypical version of biblical manuscripts. We could be dealing with a "wild" text of the bible. A "wild" text is of a form vastly different from a "standard" version. This may be true when considering the entire Reworked Pentateuch, but since 4Q158 largely consists of direct quotes from sections of Genesis, Deuteronomy, and Exodus, it by itself would not be considered a wild text.

Fragment 1-2 contains lines from Genesis 32:24-32 and Exodus 4:27-28, with extrabiblical additions. An addition to Genesis 32:30 is the exact wording Jacob received from God. Another interesting point is the tradition told in Genesis 32:32 'one does not eat from a certain portion of the thigh muscle' is transformed into a direct command from God. Fragment 4, lines 1and 2 appear to be the second half of Exodus 3:12. Lines 4-5 are a variation of Exodus 24:4-6, and the final lines are extrabiblical reflecting Gods covenant with the patriarchs.

Fragment 6 contains Exodus 20:19-21. This is expanded upon with Gods affirmation of Moses' statement to the people, and goes onto instill additional fear by holding them accountable to live up to God's commandments.

Fragment 7-8 combines Exodus 20:12-17, Deuteronomy 5:30-31, Exodus 20:22-26, and Exodus 21:1-10, with small extrabiblical additions. The 1st half of line 5 is such an addition. It is suggested by Wise, Abegg-Jr, and Cook that there may be some attempted biblical interpretation taking place. For example by mixing Exodus 20 with Deuteronomy 5 the author may have been attempting to clear up the confusing chronology surrounding the revelation at Sinai. Ancient scholars noticed that Moses went up the mountain seven times, but only explicitly descended twice. In order to correct the chronology one has to rearrange the order of the events. Certain aspects of 4Q158 seem to represent this sort of problem solving.

Fragment 10-12 closely parallels Exodus 21:32 to 22:13 with a small quantity of minor differences.

Fragment 14, an extrabiblical passage, records God speaking in the first person to a then current leader of Israel, probably Abraham or Jacob. God is providing a view into the future by revealing his intentions to desolate Egypt and promote the position of Israel for generations to come.

4Q158

Fragments 1-2

₃*Gn 32:25-30* Jacob wrestled with [a man] until [daybreak. Realizing he could not win against Jacob the man struck him on his thigh, and dislocated his hip]. They continued wrestling until early morning. [The man said, "Let me go for the day is breaking:" Jacob agreed to let go if the man ₅blessed him] The man asked him his name and Jacob told him. [The man said, "You shall no longer be called Jacob, but Israel. For you have striven with God] and humans and have prevailed." Jacob asked him his name, and he said, "Why is it you ask my name?" Then he blessed Jacob and said, " May the Lord make you fruitful, [know]ledgeable, insightful, and prevent you from sin for this day, and forever [...]" ₁₀Then the man left.

Gn 32:31-33 Jacob named the place where he saw God face to face, Penuel. As Jacob was leaving Penuel (limping because of his injury) God appeared and said, "You shall not eat [the thigh muscle that is on the hip socket." To this day the Israelites do not eat the thigh muscle] on the hip socket, [because he struck Jacob there.]

Ex 4:27- 28[God said] to Aaron, "Go [into the wilderness] to meet [Moses." Aaron went and met Moses at the mountain of God where he kissed him. Moses told] Aaron everything ₁₅God said to him. Moses said, "The Lord [has spoken] to me, saying, 'When you have brought the people out [of Egypt ...'] to go as slaves, and consider, they number thirty ....] the Lord, God [...]

Fragments 3

₁And Jacob called [...] in this earth [...] my fathers in order to enter [...]

Fragments 4

₁When you bring the people out of Egypt worship me on this mountain. So Moses built an altar there, and set up twelve pillars to represent the twelve tribes of Israel. Then he prepared a burnt offering on the altar [...Moses took half the blood and put it] ₅in bowls, *Ex 24:6* and the other half he painted on the [altar ... God said to Moses, "...] that I revealed to Abraham and to Isaac [and to Jacob ... the contract that I made] with them to be their God, and the people's [...] forever...

Fragments 6

₁ *Ex 19:20- 21* [like us, and live. Come and hear everything God tells us. Then tell us everything God said. [we will listen to you, but] don't let [God] talk to us or we [will die." Moses said, "Do not fear; for God has come to test you] [and t]o put the fear of [God in you so that you do not sin." The people stood at a distance, while Moses entered the darkness where] God was.
God said to Moses, ["I have heard what the people have said to you. They are

correct in all they have said. If] ₅they continue to fear [Me and obey all the commandments all may go well for them and their children forever! Now that you have heard] My words tell them, ['I will give them a prophet like you from their own people; This prophet will speak everything I tell him. Anyone] who does not listen [to this prophet, I will hold accountable.

Any prophet who speaks falsely in My name,] or spea[ks in the name of other gods will die. You may ask, "How will we know if a prophet speaks the LORDs words?"] If [the things a prophet says do not happen it's not the LORDs word. This prophet has spoken presumptuously, but do fear him."]

Fragments 7-8

₁*Ex 20:12- 17* Honor your [father] and your mother, [so that you may live long in the land God will give you. You shall not murder. You shall not commit adultery. You shall not steal. You shall not bear] false witness [against] your [neighbor]. You shall not desire [your] nei[ghbor's] wife, [slave, ox, donkey, or anything that belongs to your neighbor]. God said to Moses, *Dt 5:30- 31* "Tell them to [return to their tents. And with you next to Me I will tell them all the commandments, statues] and ordinances you are to teach them. They are to follow these rules in the land [I am about to give them"...]

₅The people returned to their tents, but Moses remained before *Ex 20:22-26* [God. God said, "tell the Israelites] 'that they have seen Me speak to you from heaven. They are not to make [gods of silver or gold. They need to only make an altar of earth, and sacrifice] on it your burnt offerings and offers of well being, your sheep [and oxen. Every place where I cause My named to be remembered I will come and bless you. If] you build me [an altar of stone] do not use formed stone. For by using a chisel [upon it you ruin it. Do not go up steps to My altar or your nakedness will be exposed] on it"

Fragments 10-12

₁ *Ex 21:32- 37* [If a bull kills a slave man or woman, the bull's owner is to pay the slave owner] thirty sil[ver] shekels [and the bull is to be stoned] ₂ [The owner] of an uncovered well [is responsible to compensate] the owner of any bull or ass that falls into it and return the [dead animal to the owner. If a bull kills someone else's bull, the killing bull is to be sold] with the proceeds from the sale and the dead bull [shared between both owners]. ₄ However, if the killing bull [was known to] gore [and its owner did not keep it in, the bulls would be exchanged live for dead]. ₅ If someone steals a bull or ewe and slaughters it or s[ells] it, [he is to repay the owner with five bulls for a stolen bull and four sheep for a stolen ewe. If the thief was caught during the break in] and was killed, he should not be the subject of blood retaliation. *Ex 22:1-13* If the thief was caught stealing in daylight hours he will be subjected to blood retaliation. [The thief must pay for what is stolen. If he has no worth, he will be sold for the value of the stolen property. If ₇ the stolen property is found in his possession, he will pay double if it is a live bull, ass or ewe. If a man allows his flock to graze on a field or vineyard owned by someone else, he is to repay with the produce from his own field. If he has allowed his flock to consume the entire field, he is to repay with the best of his fields or vineyards. Damages to a field as a result of fire are the responsibility of the person who started the fire. Thieves who steal property that has been entrusted in a neighbor's care are required to pay double. If the thief is not found, the person to whom the property was entrusted shall approach the house of God to decide if he is guilty of the theft. Both parties are to present their case to God, and if either party is convicted he shall pay his neighbor double. $_{12}$ [When someone has given] livestock to his neighbor for safekeeping and it dies, becomes injured or is stolen without an eye witness, judgment will be made while both parties are under oath before the Lord. If the entrusted neighbor states under oath that he did not harm or steal the owner's property, the owner is forced to agree and no repayment is required. If the entrusted neighbor was present at the time of the theft, he is required to pay damages to the owner. If the animal as torn to pieces, $_{14}$[the remains are to be shown to the owner and payment is not required.] If an animal is loaned to a friend and it dies or is torn to pieces without the owner present, the borrower is responsible for payment of damages.

Fragment 14

$_2$ [all the fl]esh and all the spirits $_3$ [...] as a blessing for the land $_4$[...] the people [...] this; Egypt shall be desolated [...] I shall create in [...] [I will rescue them from] the restraint of Egypt's power and liberate them from Egyptian control. I shall make them My people forever [and ever...I will bring them out] of Egypt. $_7$ Future generations [will settle in the] land safely for[ever ... but, I will hurl Egypt into] the heart of the sea at the deepest part [...] where they will live $_9$ [...] [bo]rders [...]

Enoch and the Watchers (4Q227)

This fragmentary manuscript is similar to portions of the book of Jubilees, an important writing of Second Temple Judaism that survived only among Christian readers and that has long been known to us from versions in Greek and Ethiopic. Among Ethiopian Christians Jubilees was so treasured that it actually became a part of the Old Testament. Fifteen fragmentary exemplars of Jubilees have turned up among the scrolls, establishing the work as one of the most common among those caches and clearly testifying to its importance for those who hid the texts. Like the Ethiopian Christians, they may have considered the book a part of the canon of Holy Writ/

In that light, the present work seems to be a retelling of Jubilees, and it may be that we should consider it an example of "rewritten Bible," the interpretive phenomenon we encounter so often in the scrolls. Surviving fragments of 4Q227 relate to Jubilees 4:17-24, but give the material in a different order. Jubilees 4:18 reports that the angels taught Enoch the calendar, which seems to be the subject of our fray. 2, 1. 1. Jubilees 4:22 says that Enoch testified against the Watchers, or fallen angels, who had taken human wives and whose progeny were the Giants (Gen. 6:1-2; cf. text 33, The Book of Giants). Our author also relates this story, in 1. 4, and apparently goes on to connect it, under the influence of Jubilees 4:23., to the judgment of the entire world.

Frag. 2 i[. . . E]noch, after we taught him 2[. . . he was with the angels of God] six full jubilees 3[. . . the la]nd, into the midst of the sons of man and he test)fied against them alI 4[. . .] and also against the watchers. And he wrote all [. . .] heaven and the ways of their hosts and [ho]ly ones 6[. . . SO th]at the ri[ghteous ones] shall not commit error [. . .]

Enoch (Hanokh, 4Q201)

One of the most important apocryphic works of the Second Temple Period is Enoch. According to the biblical narrative (Genesis 5:21-24), Enoch lived only 365 years (far less than the other patriarchs in the period before the Flood). Enoch "walked with God; then he was no more for God took him."

The original language of most of this work was, in all likelihood, Aramaic (an early Semitic language). Although the original version was lost in antiquity, portions of a Greek Text were discovered in Egypt and quotations were known from the Church Fathers. The discovery of the texts from Qumran Cave 4 has finally provided parts of the Aramaic original. In the fragment exhibited here, humankind is called on to observe how unchanging nature follows God's will.

The Book of Enoch is a pseudoepigraphal work (a work that claims to be by a biblical character). The Book of Enoch was not included in either the Hebrew or most Christian biblical canons, but could have been considered a sacred text by the sectarians. The original Aramaic version was lost until the Dead Sea fragements were discovered. Copied ca. 200-150 B.C.E.

Fragment A height 17.5 cm (6 7/8 in.), length 17.5 cm (6 7/8 in.)
Fragment B height 6.4 cm (2 1/2 in.), length 6.9 cm (2 11/16 in.)

Ena I ii

12. ...But you have changed your works,
13. [and have not done according to his command,
    and tran]sgressed against him; (and have spoken)
    haughty and harsh words, with your impure mouths,
14. [against his majesty, for your heart is hard].
    You will have no peace.

Ena I iii

13. [They (the leaders) and all ... of them took for themselves]
14. wives from all that they chose and [they began to cohabit with them and to defile themselves with them];
15. and to teach them sorcery and [spells and the cutting of roots; and to acquaint them with herbs.]
16. And they become pregnant by them and bo[re (great) giants three thousand cubits high ...]

The Book of Giants (4Q203, 1Q23, 2Q26, 4Q530-532, 6Q8)

It is fair to say that the patriarch Enoch was as well known to the ancients as he is obscure to modern Bible readers. Besides giving his age (365 years), the book of Genesis says of him only that he "walked with God," and afterward "he was not, because God had taken him" (Gen. 5:24). This exalted way of life and mysterious demise made Enoch into a figure of considerable fascination, and a cycle of legends grew up around him. Many of the legends about Enoch were collected already in ancient times in several long anthologies. The most important such anthology, and the oldest, is known simply as The Book of Enoch, comprising over one hundred chapters. It still survives in its entirety (although only in the Ethiopic language) and forms an important source for the thought of Judaism in the last few centuries B.C.E. Significantly, the remnants of several almost complete copies of The Book of Enoch in Aramaic were found among the Dead Sea Scrolls, and it is clear that whoever collected the scrolls considered it a vitally important text. All but one of the five major components of the Ethiopic anthology have turned up among the scrolls. But even more intriguing is the fact that additional, previously unknown or little-known texts about Enoch were discovered at Qumran. The most important of these is The Book of Giants.

Enoch lived before the Flood, during a time when the world, in ancient imagination, was very different. Human beings lived much longer, for one thing; Enoch's son Methuselah, for instance, attained the age of 969 years. Another difference was that angels and humans interacted freely -- so freely, in fact, that some of the angels begot children with human females. This fact is neutrally reported in Genesis (6:1-4), but other stories view this episode as the source of the corruption that made the punishing flood necessary. According to The Book of Enoch, the mingling of angel and human was actually the idea of Shernihaza, the leader of the evil angels, who lured 200 others to cohabit with women. The offspring of these unnatural unions were giants 450 feet high. The wicked angels and the giants began to oppress the human population and to teach them to do evil. For this reason God determined to imprison the angels until the final judgment and to destroy the earth with a flood.

Enoch's efforts to intercede with heaven for the fallen angels were unsuccessful (1 Enoch 6-16).

The Book of Giants retells part of this story and elaborates on the exploits of the giants, especially the two children of Shemihaza, Ohya and Hahya. Since no complete manuscript exists of Giants, its exact contents and their order remain a matter of guesswork. Most of the content of the present fragments concerns the giants' ominous dreams and Enoch's efforts to interpret them and to intercede with God on the giants' behalf. Unfortunately, little remains of the independent adventures of the giants, but it is likely that these tales were at least partially derived from ancient Near Eastern mythology. Thus the name of one of the giants is Gilgamesh, the Babylonian hero and subject of a great epic written in the third millennium B.C.E.

A summary statement of the descent of the wicked angels, bringing both knowledge and havoc. Compare Genesis 6:1-2, 4.

1Q23 Frag. 9 + 14 + 15 2[. . .] they knew the secrets of [. . .] 3[. . . si]n was great in the earth [. . .] 4[. . .] and they killed manY [. .] 5[. . . they begat] giants [. . .]

The angels exploit the fruitfulness of the earth.
4Q531 Frag. 3 2[. . . everything that the] earth produced [. . .] [. . .] the great fish [. . .] 14[. . .] the sky with all that grew [. . .] 15[. . . fruit of] the earth and all kinds of grain and all the trees [. . .] 16[. . .] beasts and reptiles . . . [al]l creeping things of the earth and they observed all [. . .] |8[. . . eve]ry harsh deed and [. . .] utterance [. . .] l9[. . .] male and female, and among humans [. . .]

The two hundred angels choose animals on which to perform unnatural acts, including, presumably, humans.
1Q23 Frag. 1 + 6 [. . . two hundred] 2donkeys, two hundred asses, two hundred . . . rams of the] 3flock, two hundred goats, two hundred [. . . beast of the] 4field from every animal, from every [bird . . .] 5[. . .] for miscegenation [. . .]

The outcome of the demonic corruption was violence, perversion, and a brood of monstrous beings. Compare Genesis 6:4.
4Q531 Frag. 2 [. . .] they defiled [. . .] 2[. . . they begot] giants and monsters [. . .] 3[. . .] they begot, and, behold, all [the earth was corrupted . . .] 4[. . .] with its blood and by the hand of [. . .] 5[giant's] which did not suffice for them and [. . .] 6[. . .] and they were seeking to devour many [. . .] 7[. . .] 8[. . .] the monsters attacked it.
4Q532 Col. 2 Frags. 1 - 6 2[. . .] flesh [. . .] 3al[l . . .] monsters [. . .] will be [. . .] 4[. . .] they would arise [. . .] lacking in true knowledge [. . .] because [. . .] 5[. . .] the earth [grew corrupt . . .] mighty [. . .] 6[. . .] they were considering [. . .] 7[. . .] from the angels upon [. . .] 8[. . .] in the end it will perish and die [. . .

] 9[. . .] they caused great corruption in the [earth . . .] [. . . this did not] suffice to [. . .] "they will be [. . .]

The giants begin to be troubled by a series of dreams and visions. Mahway, the titan son of the angel Barakel, reports the first of these dreams to his fellow giants. He sees a tablet being immersed in water. When it emerges, all but three names have been washed away. The dream evidently symbolizes the destruction of all but Noah and his sons by the Flood.

2Q26 [. . .] they drenched the tablet in the wa[ter . . .] 2[. . .] the waters went up over the [tablet . . .] 3[. . .] they lifted out the tablet from the water of [. . .]
The giant goes to the others and they discuss the dream.
4Q530 Frag.7 [. . . this vision] is for cursing and sorrow. I am the one who confessed 2[. . .] the whole group of the castaways that I shall go to [. . .] 3[. . . the spirits of the sl]ain complaining about their killers and crying out 4[. . .] that we shall die together and be made an end of [. . .] much and I will be sleeping, and bread 6[. . .] for my dwelling; the vision and also [. . .] entered into the gathering of the giants 8[. . .]
6Q8 [. . .] Ohya and he said to Mahway [. . .] 2[. . .] without trembling. Who showed you all this vision, [my] brother? 3[. . .] Barakel, my father, was with me. 4[. . .] Before Mahway had finished telling what [he had seen . . .] 5[. . . said] to him, Now I have heard wonders! If a barren woman gives birth [. . .]
4Q530 Frag. 4 3[There]upon Ohya said to Ha[hya . . .] 4[. . . to be destroyed] from upon the earth and [. . .] 5[. . . the ea]rth. When 6[. . .] they wept before [the giants . . .]
4Q530 Frag. 7 3[. . .] your strength [. . .] 4[. . .] 5Thereupon Ohya [said] to Hahya [. . .] Then he answered, It is not for 6us, but for Azaiel, for he did [. . . the children of] angels 7are the giants, and they would not let all their poved ones] be neglected [. . . we have] not been cast down; you have strength [. . .]

The giants realize the futility of fighting against the forces of heaven. The first speaker may be Gilgamesh.
4Q531 Frag. 1 3[. . . I am a] giant, and by the mighty strength of my arm and my own great strength 4[. . . any]one mortal, and I have made war against them; but I am not [. . .] able to stand against them, for my opponents 6[. . .] reside in [Heav]en, and they dwell in the holy places. And not 7[. . . they] are stronger than I. 8[. . .] of the wild beast has come, and the wild man they call [me].
9[. . .] Then Ohya said to him, I have been forced to have a dream [. . .] the sleep of my eyes [vanished], to let me see a vision. Now I know that on [. . .] 11-12[. . .] Gilgamesh [. . .]

Ohya's dream vision is of a tree that is uprooted except for three of its roots; the vision's import is the same as that of the first dream.
6Q8 Frag. 2 1three of its roots [. . .] [while] I was [watching,] there came [. . . they moved the roots into] 3this garden, all of them, and not [. . .]

Ohya tries to avoid the implications of the visions. Above he stated that it referred only to the demon Azazel; here he suggests that the destruction is for the earthly rulers alone.

4Q530 Col. 2 1concerns the death of our souls [. . .] and all his comrades, [and Oh]ya told them what Gilgamesh said to him 2[. . .] and it was said [. . .] "concerning [. . .] the leader has cursed the potentates" 3and the giants were glad at his words. Then he turned and left [. . .]

More dreams afflict the giants. The details of this vision are obscure, but it bodes ill for the giants. The dreamers speak first to the monsters, then to the giants.

Thereupon two of them had dreams 4and the sleep of their eye, fled from them, and they arose and came to [. . . and told] their dreams, and said in the assembly of [their comrades] the monsters 6[. . . In] my dream I was watching this very night 7[and there was a garden . . .] gardeners and they were watering 8[. . . two hundred trees and] large shoots came out of their root 9[. . .] all the water, and the fire burned all 10[the garden . . .] They found the giants to tell them 11[the dream . . .]

Someone suggests that Enoch be found to interpret the vision.

[. . . to Enoch] the noted scribe, and he will interpret for us 12the dream. Thereupon his fellow Ohya declared and said to the giants, 13I too had a dream this night, O giants, and, behold, the Ruler of Heaven came down to earth 14[. . .] and such is the end of the dream. [Thereupon] all th e giants [and monsters! grew afraid 15and called Mahway. He came to them and the giants pleaded with him and sent him to Enoch 16[the noted scribe]. They said to him, Go [. . .] to you that 17[. . .] you have heard his voice. And he said to him, He will [. . . and] interpret the dreams [. . .] Col. 3 3[. . .] how long the giants have to live. [. . .]

After a cosmic journey Mahway comes to Enoch and makes his request.

[. . . he mounted up in the air] 4like strong winds, and flew with his hands like ea[gles . . . he left behind] 5the inhabited world and passed over Desolation, the great desert [. . .] 6and Enoch saw him and hailed him, and Mahway said to him [. . .] 7hither and thither a second time to Mahway [. . . The giants awaig 8your words, and all the monsters of the earth. If [. . .] has been carried [. . .] 9from the days of [. . .] their [. . .] and they will be added [. . .] 10[. . .] we would know from you their meaning [. . .] 11[. . . two hundred tr]ees that from heaven [came down . . .]

Enoch sends back a tablet with its grim message of judgment, but with hope for repentance.

4Q530 Frag. 2 The scribe [Enoch . . .] 2[. . .] 3a copy of the second tablet that [Epoch] se[nt . . .] 4in the very handwriting of Enoch the noted scribe [. . . In the name of God the great] 5and holy one, to Shemihaza and all [his companions . . .] 6Iet it be known to you that not [. . .] 7and the things you have done, and that your wives [. . .] 8they and their sons and the wives of [their sons . . .] 9by your licentiousness on the earth, and there has been upon you [. . . and the land is crying out] 10and complaining about you and the deeds of your children [. . .] 11the harm that you have done to it. [. . .] 12until Raphael arrives, behold, destruction [is coming, a great flood, and it will destroy all living things] 13and whatever is in the deserts and the seas. And the meaning of the matter [. . .] 14upon you for evil. But now, loosen the bonds bi[nding you to evil . . .] 15and pray.

A fragment apparently detailing a vision that Enoch saw.
4Q531 Frag. 7 3[. . . great fear] seized me and I fell on my face; I heard his voice [. . .] 4[. . .] he dwelt among human beings but he did not learn from them [. . .]

Exhortation based on the Flood (4QFloodAp=4Q370)

This portion of the Dead Sea Scrolls, 4Q370, tells much the same story that was told in Genesis Chapter 6. It does not tell the story of Noah and his ark, but rather explains how God felt that the people of the earth were evil, and therefore flooded the earth. Also, this portion of the Dead Sea Scrolls is similar to the story in the Bible when it tells of God's promise to not flood the earth again. Column I of the text gives a description of the Flood, while the second column tells the intention of the story. The application of the text is that at Creation the Lord gave so much t the people that it caused corruption.

Col. I

₁ And the Lord covered the land with fruits and gave them plenty of food and made every living thing content with the fruit. "May everyone who does as I ask be filled with food and be satisfied," said the Lord, "and show devotion to my [holy] name." "But now they have done things that I believe are evil, "God said. And they went against what God asked through their actions. And God judged them according to their actions and their thoughts of the [immoral] tendency of their evil hearts and thundered against them with through his power. And the entire earth shook, and the waters overflowed from the gorges; all the entrance gates of the heavens opened up and the abysses overflowed with strong waters; ₅ and the entrance gates of the heavens poured rain. And they were destroyed by the flood.[...] everyone died in the waters...[...] This is why everything that was on land [disappeared,] and men, the [animals and all the] birds, everything with wings [died.] Not even the strongest escaped. [...]...And God made [a contract] and put the rainbow [in the clouds] to remember the contract he made with the people [...and never again will] a flood [come] for [destruction, or] will the chaos of the waters be opened. [...] they made, and clouds [...] for (the) waters [...] ₁₀ [...]

Col. II

₁ Because they felt guilt from their sin, they will ask [...] God will justify [...] and he will purify them of their sins [...] their evil from their knowledge [...] ₅ They grow, but their days are like a shadow [...] and he will always be caring [...] God's marvels; remember the mira[cles of the Lord...] due to his fear and your awe of him your heart will celebrate because he is good [...] those who support you. Do not disobey what [God] asks you to do...

Exodus 4Q22 (paleo Exod^m), Column 1

Scroll fragment of Exodus: 6:25-7:19. The remains of all Biblical books (except for Esther) were found in the Dead Sea Caves. This fragment is written in palaeo-Hebrew script, recently redacted to c 100 BCE.

2   These are that Aaron and Moses to whom
3   the Lord said, Bring out the children of Israel from the land of Egypt according
4   to their armies. These are they which spoke to Pharaoh king of Egypt to bring out the children of Israel.
5   from Egypt: these are that Moses and Aaron. And it came to pass on the day when the Lord spoke
6   unto Moses in the Land of Egypt,
7   That the Lord spoke unto Moses, saying, I am the Lord: speak thou unto Pharaoh king of Egypt
8   all that I say unto thee. And Moses said before the Lord,
9   Behold, I am of uncircumcised lips, and how shall Pharaoh hearken unto me?
10   And the Lord said unto Moses, See, I have made thee a god to Pharaoh: and Aaron thy brother shall be
11   thy prophet. Thou shalt speak all that I command thee: and Aaron thy brother shall speak unto
12   Pharaoh, that he send the children of Israel out of his land. And I will harden Pharaoh's heart
13   and multiply my signs and my wonders in the land of Egypt.

Leviticus (Va-Yikrah, 11Q1 PaleoLev)

This scroll was discovered in 1956, when a group of Ta`amireh Bedouin happened on Cave 11, but it was first unrolled fourteen years later, at the Israel Museum in Jerusalem. Inscribed in the scroll are parts of the final chapters (22-27) of Leviticus, the third book in the Pentateuch, which expounds laws of sacrifice, atonement, and holiness. This is the lowermost portion (approximately one-fifth of the original height) of the final six columns of the original manuscript. Eighteen small fragments also belong to this scroll. The additional fragments of this manuscript are from preceding chapters: Lev. 4, 10, 11, 13, 14, 16, 18-22.

The Leviticus Scroll was written in an ancient Hebrew script often referred to as paleo-Hebrew. The almost uniform direction of the downstrokes, sloping to the left, indicates an experienced, rapid, and rhythmic hand of a single scribe. The text was penned on the grain side of a sheep skin. Both vertical and horizontal lines were drawn. The vertical lines aligned the columns and margins; the horizontal lines served as guidelines from which the scribe suspended his letters. Dots served as word-spacers. Copied late second century - early first century B.C.E. Height 10.9 cm (4 1/4 in.), length 100.2 cm (39 1/2 in.)

11Q1(PaleoLev)

Lev. 23:22-29

1. (22)[...edges of your field, or] gather [the gleanings of your harvest; you shall leave them for the poor and the stranger; I the LO]RD [am]
2. your God.
3. (23)The LORD spoke to Moses saying: (24)Speak to the Israelite people thus: In the seventh month
4. on the first day of the month, you shall observe complete rest, a sacred occasion commemorated with load blasts.
5. (25)You shall not work at your occupations; and you shall bring an offering by fire to the LORD.
6. (26)The LORD spoke to Moses saying: (27)Mark, the tenth day of this seventh month is the Day
7. of Atonement. It shall be a sacred occasion for you: you shall practice self-denial, and you shall bring an offering
8. by fire to the LORD; (28)you shall do no work throughout that day. For

[it is a Day of Atonement on which] expiation is made on your behalf [before the LO]RD your God. (29)Indeed, any person who

Joshua Apocryphon (4Q522)

This text contains an assortment of geographical locales and place names that may go back to the period of Joshua or reflect some display or schema connected to the Davidic period. Because of the reference to Eleazar, the high priest often associated with Joshua's activities, we have called it a Joshua Apocryphon, though the text, as it has been preserved in the second column, clearly focuses on the figure of David, his activities, his conquests, his kingship and, in particular, his building of the Temple. For those who would refer to this literature as sectarian, the nationalist implications of texts such as this are important, as is the provocative allusion to 'the sons of Satan' in Line 5. This parallels similar references elsewhere in the corpus to 'sons of Belial', and their variation in the 'Mastemoth' / 'sons of Darkness' allusions we have encountered above and will encounter further below.

Text

Fragment 1

Column 1

(1)... and En Qeber and... (2)... Valley, and Bet Zippor, with (3)... all the Valley of Mozza (4)... and Heikhal -Yezed(?) and Yapur and (5)... and Mini and En Kober (6)... Garim and Hedita and Oshel (7)... which (8)... and Ashkalon... (9)... [G]alil, and the two... and the Sharon (10)... Judah, and Beer Sheba, and Baalot (11)... and Qeilah and Adullam and (12)... Gezer and Thamni and Gamzon and (13)... Hiqqar and Qittar and Ephronim and Shakkoth (14)... Bet Horon, the lower and the upper, and (15)... and the Upper and the Lower Gilat

Column 2

(1)... to establish there the... (2) the times, for a son is about to be born to Jesse, son of Perez, son of Ju[dah...] (3) (He shall capture) the mountain (literally, rock) of Zion, and he will dispossess from there all the Amorites... (4) to build the House for the Lord, the *God* of Israel. Gold and silver... (5) cedars and cypress will he bring from Lebanon to build it, and the sons of Satan... (6) he will do priestly service there and a man...your... (7) from the... And the Lord will establish David securely... (8) [He]aven will dwell with him forever. But now, the Amorites are there, and the Canaanites... (9) dwell where the Hittites (do), none of whom have I sought... (10) from you. And the Shilonite, and the... I have given him as a servant... (11) And now, let us establish... far from... (12) Eleazar... forever, from the House (13)... army...

A Biblical Chronology (4Q559)

This work attempted to determine the chronology of the people and, in some cases, the events of the Bible. This enterprise was important not only for its intrinsic interest, but also to those who wanted to locate the present in the flow of time toward the Messianic era. They could then predict when the Messiah would come, and when other predictions of the prophets would find fulfilment. Interest in such 'chronomessianism' was great in the period of the Scrolls.

Text

Column 1

(Fragments 1 and 3) (1) [...and Terah was (2) seventy years old when he fathered Abraham; and Abraham (3) was] ninety-nine years old [when he fathered Isaac; (4) and Isaac was [sixty years old when he fathered Jacob; and Jacob (5) was] sixty-fi[ve year]s old [when he fathered Levi... (7) and Levi was thirty-] five [year]s old when he fa[thered Kohath; and Kohath was (8) twenty-] nine [year]s old when he fathered Am[r]am; Am[ram was (9) one hundred and ten years old when he fathered] Aaron, and Aar[on] went out from Egy[pt. (10) (The total of) all] these [years:] eleven thousand, five hundred and thirty-six...

Column 2

(2)... from the lan[d of Egypt...] (3) ye[ars (?)...] (4) the [Jo]rdan... (5) [in...] thirtyfive (or more) [years;] in Gilgal... ye[ars... (6) in Timnath-Sera]h twenty years; and from the time that [Joshua] died... (7) Cush-Rishathaim the king of [Aram-Naharain] (8) eight [year]s; Othniel the s[on of Kenaz... (9) eighty years;] Eglon the king of Moab eighteen] ye[ars; (10) Eh]ud the son of Gera, eighty years; Sham[gar the son of Anath,]

Hur and Miriam (4Q544)

This Aramaic text is difficult to characterize because it is so fragmentary, but it appears to be concerned with genealogies of characters in the Book of Exodus, particularly Hur. If Line 8 is taken with Line 9, the text would appear to make a connection between the Judean hero Hur and Miriam, the sister of Moses, although none is anywhere made explicit in the Bible. According to Josephus, Hur was Miriam's husband (Ant. 3.54), and the tradition being signalled here seems to bear this out. Rabbinic literature, identifying Ephrath and Miriam, would have him as her son (Targum to 1 Chr. 2:19 and 4:4). Still, that there is tradition about a relationship between Hur and Miriam is not to be gainsaid.

Text

Fragment 1

(1) [th]at he ate, he and his son[s... (2) [and] her [hu]sband [slept] the eternal sleep... (3) upon him, and they found hi[m... (4) his sons and the sons of h[is] brother... (5) they dwelt temporarily (?)... (6) he departed to his Eternal home... (8) ten. And with Miriam he became the father of Ab[(?; name incomplete and uncertain)... (9) and Sitri. Then Hur took as wife... (10) And with her he became the father of Ur and Aar[on... (11) with her four (forty?) sons...

Pseudo-Jubilees (4Q227)

The preserved portions of these two fragments appear to contain information about Enoch similar to the Book of Jubilees 4:17-24. A good deal of the interest centering around Enoch, as we have noted, was connected with his assumption alive into Heaven and the mysterious allusion in Genesis to 'walking with *God*'. The use, too, of darcheizevam ('the paths of their hosts') in 2. 5, meaning the fixed trajectories of the heavenly bodies is interesting too, as a heavenly parallel to some of the more common, earthly adumbrations of this terminology for the Community, such as the Darchei-Zedek, Darchei-Emet, Darchei-{c}Or ('the Ways of Righteousness', 'Truth', 'Light'), etc.., as opposed to 'the Ways of Darkness', 'Evil', 'Lying', 'uncleanness', 'abomination', 'fornication' and the like. By implication, these too are fixed, at least in their positive sense, by the Law.

Text

Fragment 1

(1)... all the Righteous... (2) before Moses... (3) all the days of... (6) the years of... will be lengthened...

Fragment 2

(1) [E]noch after we instructed him (2)... six jubilees of years (3)... [ea]rth among all mankind and he witnessed against them all (4)... and also against the Watchers. And he wrote down all the (5)... [the] heavens and all the paths of their hosts (i.e., the heavenly bodies), all [the months (6)... [in which the Rig[hteous] have not erred...

Aramaic Tobit (4Q 196)

For Jews and Protestant Christians the book of Tobit is outside the canon of the Bible, being counted among the Apocrypha. Catholics, however, along with the Greek and Russian Orthodox branches of Christianity, regard the book as part of the Bible in the sense that it is 'Deuterocanonical'. Although scholars for the most part believed that Tobit had originally been written in either Hebrew or Aramaic, the Semitic original version was long lost.

The book's primary witnesses were two rather different Greek versions (one 'short' version and one 'long'). Thus the significance of the Qumran caches: they include portions of four Aramaic manuscripts of the book, together with one Hebrew manuscript. All of these manuscripts support the 'long' version of Tobit known from the Greek.

Text

Fragment 1

(1) [one o]f the Ninevites (went) and informed the kin[g that] I was bury[ing them, so] I hid myself. When I discovered [that] he knew about me, (2) [and that I was being sought to be killed, I was frightened and I fled. [Then a]ll [that] I owned [was confiscated,] nothing was left to me. (There was not) one thi[ng] (3) [that they did

not take] to the [king's storehouse,] ex[cept Hannah my wife and T[ob]iah my son. It was not fo[rty-] (4) [five days before two of] his [sons killed him (i.e., the king). They fled to the mountains of Ararat, and [Esarhaddon his son] ruled (5) [in his place. He em]powered Ahiqar the son of Anael my brother over all the ac[counting] (6) [of his kingdom. He also installed Ahiqar as chief of the (sacred) treasury. He was in charge of [a]ll of the king's funds. Aqihar interceded (7) on my behalf, so that I returned to Nineveh. Ahi]qar my 'brother' was the chief cupbearer and the official in charge of the signet rings and the treasurer (8) [and] the accountant for Sennacherib, king of Assyria, while Esarhaddon appointed him second (only) to himself. For [he] (9) was my nep[hew], from the house of my father and my family. In the days of Esarhaddon the [ki]ng, when I returned (10) to my home and Hannah my wife was returned to me, [along with] Tobiah my son, on the day of the festival of We[eks, I had] (11) a fine banquet. I reclined to [ea]t, and they set the ta[b]le before me. I saw the many delicacies that they brought, (12) and [I sa]id [to To]biah my son, 'My son, go and bring anyone [whom you] may find among [our] brothers (13)... my son, go and get (such), that he may come and eat [together with me...]

Stories from the Persian Court (4Q550)

Apparently these stories concerned the adventures of Jews at the court of the Persian kings. The use of the word 'Jew' (Yehudi) in 6.3 is, therefore, probably one of the oldest usages of this term - it is not commonly found in other genres of Qumran literature, which tend either to speak in terms of classical archetypes or to archaize (cf. the allusion to Beit-Yehudah/ 'House of Judah' in the Habakkuk Pesher, viii). We will find it in other interesting texts below, and apparently the usage was already becoming common even in Palestine. One also begins to encounter it on the coins of the Maccabeans also discussed below.

Text

Column 2 (or later)

(1) a man, unless the king knows; indeed, there exists... (2) and the witness shall not perish. They have believed what is upright... (3) 0 king, Fratervana the son of... has... (4) there fell upon him fear of the (contents of) the archi[ves...] (5) the foundations of the king that you shall spe[ak] and be given... (6) my house [and] my [po]ssessions for everything that... (7) shall you be able to take up your father's occupation?...

Column 3 (or later)

(1) the foundations of the king that you shall speak to the prince (?)... (2) Fratervana [your] father from the day that he took up his occupation before the

king... (3) for him he did honestly and... before him... (4) and he said the foundations of... (7) pea[ce]...

Column 4 (or later)

(1) [they used] to listen to Fratervana your father... (2) to the servants of the royal wardrobe in [al]l... to do (3) the business of the king in all that he recei[ved...] in that very year (4) the king's patience... his fat[h]er... before him; among (5) the books was found a cer[tain] scroll [sea]led by seven seals (impressed by) the signet ring of Darius his father. The matter (6)... '[Dar]ius the king to the servants of the kingdom of a[ll] the [e]arth: Peace.' It was opened and read. (The following) was found written in it: 'Darius the king (7) [wrote to all] kings after myself, to the servants of the kingdom: Pe[ac]e. Let it be known to you that all oppression and falsehood...

Column 6 (or later)

(1)... for you know... for the sins of my fathers (2) that they sinned aforetime and... and I followed after... (3) a Jew from among the k[ing]'s officials stood in front of him and... [the] good [ma]n. (4) The good man did... What shall I do with you? You know [that] it is poss[ible] (5) for a man like [you to hasten (?) everything. A man] of your household (once) stood where you (now) stand... (6) Command me (to do) any[thi]ng that you wa[n]t, and when you have [spo]ken, I will bury you in... (7) he dwells in all. It is possible that he will bring in my service be[fore... and everything that...

Column 8 (or later)

(1) the highest (*God*) whom you fear and serve, he is ruler over the [ear]th. It is easy for him to [d]o anything that he desires, (2) [and] anyone who speaks an [e]vil word against Bagasri... shall be killed, because there is no... (3) Good forever... that he saw... two. And he said, 'Let the king write... (4) he saw... to the k[ing...] them in the great ro[y]al court... (5) and after (the story of) Badgers, they read in thi[s] book... (6) Evil, his Evil shall return upon his...

Column 9 (or later)

(1) [the] k[ing]'s decree... they went... (2) [to] writ[e...] he went... in the clothing... (3) a gold[en] crown [weighing one hundred and fi[f]ty. He went... (4) apart from him... he went and sai[d...] (5) (he returned the?) [si]lver and [g]old and [possession]s that [bel]ong to Bagose in a double measure... (6) he entered the king's court in the name of Bagasri... (7) [ki]lled. Then [B]agasri entered the king's co[ur]t... (8) the chief butler answered and said, 'Bagasri, Bagasri, from...'

Fernando Klein

Legal and Ritual Texts

Phylacteries

The command *"And thou shalt bind them for a sign upon thy hand, and they shall be for frontlets between thine eyes"* (Deut. 6:8) was practiced by Jews from early times. In the Second Temple period the sages established that *tefillin* (phylacteries; amulets in Greek) would include four scriptural passages inscribed on parchment placed in box-like containers made of black leather. One of the phylacteries was worn one on the left arm and the other on the forehead. These served *"as a sign upon your hand and as a symbol on your forehead that with a mighty hand the Lord freed us from Egypt"* (Exodus 13:9, 16). The Dead Sea region has now yielded the earliest phylactery remains, both of the leather containers and the inscribed strips of parchment. As a rule, phylacteries include the same four selections, two from the book of Exodus (Exod. 13:1-10; 13:11-16) and two from Deuteronomy (Deut. 6:4-9; 11:13-21). The scriptural verses were penned in clear minuscule characters on the elongated writing material, which was folded over to fit the minute compartments stamped into the containers.

Exod. 13:1-3

(1) And spoke the Lord to Moses saying,
(2) "Consecrate to Me every first-born the first issue of every womb of the Israelites, man and beast is Mine."
(3) And Moses said to the people, "Remember this day on which you went (free) from Egypt, the house of bondage, how with a mighty hand the Lord freed you from it; no leavened bread shall be eater.
(4) This day

Ritual Purity Laws (4QTohorota=4Q274)

The ritual purity laws are found in Leviticus 13-15 in the Bible. It addresses diseases and discharges that cause contamination such as: leprosy, seminal discharge, discharge of blood, the Niddah (a mentraunt woman), and contact with corpses. All of these impurities are alike for one reason: people with these impurities were ostracized from the towns and from the holy temple. The destruction of the temple in Jerusalem in 586 BC and 70 CE helped to remove the "justification of the laws of impurity" (Biale 147). After the destruction of the temple, the laws concerning a Niddah became less severe. Before the temples were destroyed, a woman who had her menstrual cycle was not allowed to touch anyone until her menstruation was over. After the destruction the laws were not as strict, and it came to be that the only thing a Niddah was not allowed to do was have sex with her husband. There are some difficulties that one will run into when researching these laws. Some Texts will show three fragments while others will show only two. There is no information on the laws themselves, like what language they are in. The information about the laws is very scarce and is mostly on what the laws contain.

4Q274

Fragment 1 col.1:

$_1$He will begin by not] casting his lot [?for priestly service?]. He will lie down in the bed of trouble, and reside in a house of grief. He will live away from the pure, $_2$with all the unclean at the distance of twelve cubits . He will live to the northeast of any habitation at the same distance. $_3$Anyone who has a discharge, will bathe and wash his clothes and afterwards he may eat. For it says (Lev. 13:45) "unclean, unclean" $_4$they will shout all the days of their discharge. And she who is discharging blood, for seven days she may not touch the man who has a discharge or any of the objects that he uses. $_5$Also for any of the objects he has laid on or sat on. And if she touches anything she will wash her clothes and bathe and afterwards she may eat. In no way may she mingle during her seven $_6$days so she does not contaminate the camps of the holy ones of Israel. She may not touch any woman who has had a discharge of blood for several days. $_7$And the one who is counting their seven days, whether they are male or female may not touch ... during the start of her period, unless she is clean from her menstruation. For the blood $_8$from menstruation is considered a discharge for anyone who touches it. And if a flow of semen is discharged, it is a misfortune. And he will be unclean... and anyone who touches $_9$any of these unclean people, they will not eat during their seven days of their impurity, just like the person who is unclean through contact with a corpse. And they will bathe and wash and then...

Fragment 1 col.2:

$_1$...which they sprinkle on themselves the first time, and they will bathe and wash before $_2$... they will immerse themselves the seventh time on the Sabbath day. $_3$They may not touch the pure food until they change their clothes $_4$... anything that touches the discharge of semen, whether it is a person or an object, they will immerse, and the one who carries it $_5$will immerse... and they will immerse the garment which is on them and the object which they carry $_6$... And if there is a man in the camp whose hands or feet has not reached...$_7$ the garment which has not touched it. Only they may not touch their food. And the one who touches it,$_8$ will immerse... they will live alone. If they have not touched it, was their clothes in water and if $_9$... and they will wash. And concerning all holy things, they will wash in water...

Fragment 2 col.1:

$_1$...when God reveals the apple of his eye and he calls out $_2$... and every statute...$_3$ who eats... $_4$not... $_5$it is their flesh and it is unclean $_6$... their drink and they may not eat the pure food and all $_7$... after they are pressed and their juices run out, no one may eat them $_8$... if the unclean person touches them and also the greens...$_9$ or boiled cucumber, and the person who waters...

Fragment 2 col.2:

$_1$...they are unclean. The...$_3$ Anything which has a seal... $_4$they will leave all the greens for the person who is cleansed...$_5$ from the moisture of the dew, they may eat, but if not...$_6$ in the middle of the water unless a person...$_7$ the land, if they come against it...$_8$ the rain on it, and if the... touches it... $_9$on the field in all its measures in respect to the season of the year...$_{10}$ any clay object that will fall in it... and any $_{11}$that are clean in its middle... and every $_{12}$drink that they will drink...

A Baptismal Liturgy (4Q414)

4Q414

F.2 Col.1
(... And he shall) say (in response)"*Blessed (are You, ...)   The unclean for the festivals of (...) Your (...) and to make atonement for us (...to be) pure before you (...) in every matter (...) to purify oneself prior to (...) You made us (....)*

F.2 + 3. Col.2
And you shall cleanse him for Your holy statutes (..) for the first, the third and the sixth (...) in the truth of Your covenant (...) to cleanse oneself from uncleanness (...) and then he shall enter the water (...) And he shall say in response "*Blessed are You (...) for from what comes out of Your mouth (...) men of impurity (...)*

F.10
Soul (...) he is (...) to Yourself as a pure people (...) And I also (...) the day which (...) in the times of purity (...) the *Yahad*. In Israel's pure food (...) and they shall dwell (...). And it will happen on that day (...) a female and she will give thanks (...)

F.12
For You made me (...) Your will is that we cleanse ourselves before (...) and he established for himself a statute of atonement (...) and to be in righteous purity and he shall bathe in water and sprinkle upon (...) (...) And then they return from the water (...) cleansing His people in the waters of bathing (...) second time upon his station. And he shall say in response : "*Blessed are You (...) (...) Your purification in Your glory (...) (...) eternally. And today (...).*

Acts of Torah (4Q394)

From a unique document in letter format, which outlines the religious laws peculiar to the sect, and in opposition to the law practiced by the Temple in Jerusalem. This copy in Herodian script dates from the late 1st Century BCE to early 1st Century CE.

4Q394(MMT), Column 4

52   and concerning the deaf who have not heard the laws and the judgements and the purity regulations, and have not
53   heard the ordinances of Israel, since he who has not seen or heard
54   does not know how to obey (the Law): nevertheless they have access to the sacred food.
55   And concerning liquid streams: we are of the opinion that they are not
56   pure, and that these streams do not act as a separative between impure
57   and pure (liquids). For the liquid of streams and (that) of (the vessel) which receives them are alike, (being)
58   a single liquid. And one must not let dogs enter the holy camp, since they
59   may eat some of the bones of the sanctuary while the flesh is (still) on them. For
60   Jerusalem is the camp of holiness, and is the place
61   which He has chosen from among all the tribes of Israel. For Jerusalem is the capital of the
62   camps of Israel. And concerning (the fruits of) the trees planted

Various Laws (4Q159)

Frag. 1, col II + 9 (= 4Q513 1-2 ?)
1   [...] Not [...]
2   [...Isra]el its ini[quiti]es and to atone for all th[eir] sins. [...]
3   [... And if] someone [ma]kes it into a threshing floor or a press, whoever comes to the threshing floor [or to the press, ...]
4   whoever in Israel owns nothing, that person can eat some and gather for himself; but for [his] house[hold he is not to gather.]
5   (In) the field he may eat it for himself, but is not to bring it to his house to store it [...]
6   [Concer]ning [the ransom:] the money of valuation which one gives a ransom for his own person will be half [a shekel,]
7   only once will he give it in all his days. The shekel comprises twenty geras in the she[kel of the temple.]
8   for the six hundred thousand; one hundred talents; for the third, half talent, [...]
9   and for the fifty, half a mina, twenty-five shekels. The total [...]
10   mina. [...t]hree for ten minas [...]
11   [...five (shekels) of silver (make) the tenth part of [a mina ...]
12   [... the she]kel fo the temple; ha[lf a shekel ...]
13   [...] the ephah and the bath are of the s[ame] size [...]
14   [... t]hree tenths [...]
15   [...] *Blank* [...]
16   [... u]pon the people, and upon [their] g[ar]ments [...]

17 [... I]srael, Mos[es] burnt [...]

Frags. 2-4

1   And if [... to a] foreigner or the descendant of a [foreign] fam[ily ...]
2   in the presence of Is[rael.] They are [no]t to serve gentiles; with...[... from the land of]
3   Egypt and command them not to be sold for the price of a slave. and [te]n men
4   and two priests, and they shall be judged by these twelve. [... In every]
5   capital offence in Israel their authority should be consulted, and whoever disobeys [...]
6   he will be executed, for the acted presumptuously. A woman is not to wear the clothes of a male; every[one ... and he is not]
7   to put on a woman's cloak, and he is not to dress in a woman's tunic, for it is an [ab]omination. [...] *Blank* [...]
8   In the case where a man slanders a maiden of Israel, if he says it at the [moment] of taking her, they shall examine her
9   regarding (her) trustworthiness. If he has not lied about her, she shall be put to death; but if he has testified [false]ly against her, they are to fine him two minas [and he is not]
10 to divorce her for all the days (of his life), Anyone who [...] to do ... [...]

Frag. 5

1   [... be]fore God, and they died. The interpretation of [...]
2   [...] *Blank* The sons of Lev[i ...]
3   [...] in judgment. And what he sa[ys ...]
4   [...] when Moses took the [...]
5   [...] they will go out there. Interpretation of the matter: [...]
6   [... to ex]pound the Law in distress and [...]
7   [... a]s Moses said [...]
8   [...] all [...]

Songs of Sabbath Sacrifice (Shirot ʿOlat ha-Shabbat, 4Q403 ShirShabbd)

The Songs of the Sabbath Sacrifice, also known as the "Angelic Liturgy," is a liturgical work composed of thirteen separate sections, one for each of the first thirteen Sabbaths of the year. The songs evoke angelic praise and elaborate on angelic priesthood, the heavenly temple, and the Sabbath worship in that temple.

The headings of the various songs may reflect the solar calendar. Although the songs bear no explicit indication of their source, the phraseology and terminology of the texts are very similar to those of other Qumran works.

Eight manuscripts of this work were found in Qumran Cave 4 (4Q400 through 407) and one in Cave 11, dating from the late Hasmonean and Herodian periods. One manuscript of the Songs of the Sabbath Sacrifice was found at Masada, a Zealot fortress. Copied mid-first century B.C.E. Height 18 cm (7 in.), length 19 cm (7 1/2 in.)

4Q403(ShirShabbd)

30. By the instructor. Song of the sacrifice of the seventh
    Sabbath on the sixteenth of the month. Praise the God of
    the lofty heights, O you lofty ones among all the

31. elim of knowledge. Let the holiest of the godlike ones
    sanctify the King of glory who sanctifies by holiness
    all His holy ones. O you chiefs of the praises of

32. all the godlike beings, praise the splendidly
    [pr]aiseworthy God. For in the splendor of praise
    is the glory of His realm. From it (comes) the praises
    of all

33. the godlike ones together with the splendor of all [His]
    maj[esty. And] exalt his exaltedness to exalted heaven,
    you most godlike ones of the lofty elim,
    and (exalt) His glorious divinity above

34. all the lofty heights. For H[e is God of gods] of all
    the chiefs of the heights of heaven and King
    of ki[ngs] of all the eternal councils.
    (by the intention of)

35. (His knowledge) At the words of His mouth come into
    being [all the lofty angels]; at the utterance
    of His lips all the eternal spirits; [by the in]tention
    of His knowledge all His creatures

36. in their undertakings. Sing with joy, you who rejoice
    [in His knowledge with] rejoicing among the wondrous
    godlike beings. And chant His glory with the
    tongue of all who chant with knowledge;
    and (chant) His wonderful songs of joy

37. with the mouth of all who chant [of Him. For He is]
    God of all who rejoice {in knowledge}
    forever and Judge in His power of all the
    spirits of understanding.

Some Torah Precepts (Miqsat Ma`ase ha-Torah, 4Q396 MMTc)

This scroll, apparently in the form of a letter, is unique in language, style, and content. Using linguistic and theological analysis, the original text has been dated as one of the earliest works of the Qumran sect. This sectarian polemical document, of which six incomplete manuscripts have been discovered, is commonly referred to as MMT, an abbreviation of its Hebrew name, Miqsat Ma`ase ha-Torah. Together the six fragments provide a composite text of about 130 lines, which probably cover about two-thirds of the original. The initial part of the text is completely missing.

Apparently it consisted of four sections: (1) the opening formula, now lost; (2) a calendar of 364 days; (3) a list of more than twenty rulings in religious law (Halakhot), most of which are peculiar to the sect; and (4) an epilogue that deals with the separation of the sect from the multitude of the people and attempts to persuade the addressee to adopt the sect's legal views. The "halakhot," or religious laws, form the core of the letter; the remainder of the text is merely the framework. The calendar, although a separate section, was probably also related to the sphere of "halakhah." These "halakhot" deal chiefly with the Temple and its ritual. The author states that disagreement on these matters caused the sect to secede from Israel. Late first century B.C.E.-early first century C.E.

Fragment A: height 8 cm (3 1/8 in.) length 12.9 cm (5 in.)
Fragment B: height 4.3 cm (1 11/16 in.) length 7 cm (2 3/4 in.)
Fragment C: height 9.1 cm (3 9/16 in.) length 17.4 cm (6 7/8in.)

4Q396(MMTc)

1. until sunset on the eighth day. And concerning [the impurity] of
2. the [dead] person we are of the opinion that every bone, whether it
3. has its flesh on it or not--should be (treated) according to the law of the dead or the slain.
4. And concerning the mixed marriages that are being performed among the people, and they are sons of holy [seed],
5. as is written, Israel is holy. And concerning his (Israel's) [clean] animal
6. it is written that one must not let it mate with another species, and concerning his clothes [it is written that they should not]
7. be of mixed stuff; and one must not sow his field and vineyard with mixed species.
8. Because they (Israel) are holy, and the sons of Aaron are [most holy.]
9. But you know that some of the priests and [the laity intermingle]
10. [And they] adhere to each other and pollute the holy seed as well as their (i.e. the priests') own [seed] with corrupt women. Since [the sons of Aaron should...]

Tehillim (11Qp1)

This impressive scroll is a collection of psalms and hymns, comprising parts of forty-one biblical psalms (chiefly form chapters 101-50), in non-canonical sequence and with variations in detail. It also presents previously unknown hymns, as well as a prose passage about the psalms composed by King David.

One of the longer texts to be found at Qumran, the manuscript was found in 1956 in Cave 11 and unrolled in 1961. Its surface is the thickest of any of the scrolls-Äit may be of calfskin rather than sheepskin, which was the more common writing material at Qumran. The script is on the grain side of the skin. The scroll contains twenty-eight incomplete columns of text, six of which are displayed here (cols. 14-19). Each of the preserved columns contains fourteen to seventeen lines; it is clear that six to seven lines are lacking at the bottom of each column.

The scroll's script is of fine quality, with the letters carefully drawn in the Jewish book-hand style of the Herodian period. The Tetragrammaton (the four-letter divine name), however, is written in the paleo-Hebrew script. Copied ca. 30 - 50 C.E. Height 18.5 cm (7 1/4 in.), length 86 cm (33 3/4 in.)

Column 19:

1. Surely a maggot cannot praise thee nor a grave worm recount thy loving-kindness.
2. But the living can praise thee, even those who stumble can laud thee. In revealing
3. thy kindness to them and by thy righteousness thou dost enlighten them. For in thy hand is the soul of every
4. living thing; the breath of all flesh hast thou given. Deal with us, O LORD,
5. according to thy goodness, according to thy great mercy, and according to thy many righteous deeds. The LORD
6. has heeded the voice of those who love his name and has not deprived them of his loving-kindness.
7. Blessed be the LORD, who executes righteous deeds, crowning his saints
8. with loving-kindness and mercy. My soul cries out to praise thy name, to sing high praises
9. for thy loving deeds, to proclaim thy faithfulness--of praise of thee there is no end. Near death
10. was I for my sins, and my iniquities have sold me to the grave; but thou didst save me,
11. O LORD, according to thy great mercy, and according to thy many righteous deeds. Indeed have I
12. loved thy name, and in thy protection have I found refuge. When I remember thy might my heart
13. is brave, and upon thy mercies do I lean. Forgive my sin, O LORD,

14. and purify me from my iniquity. Vouchsafe me a spirit of faith and knowledge, and let me not be dishonored
15. in ruin. Let not Satan rule over me, nor an unclean spirit; neither let pain nor the evil
16. inclination take possession of my bones. For thou, O LORD, art my praise, and in thee do I hope
17. all the day. Let my brothers rejoice with me and the house of my father, who are astonished by the graciousness...
18. [ ] For e[ver] I will rejoice in thee.

The First Letter on Works Reckoned As Righteousness (4Q394/8)

This text is of the most crucial importance for evaluating the Qumran community, mindset and historical development. Parts of it have been talked about, written about and known about for over three decades. Particularly in the last decade, parts have circulated in various forms, some under the by now popular code 'MMT'. In turn, this is often incorrectly spoken of as 'some words of the Torah'.

This title would only be appropriate to the First Letter, but the allusion on which it is based actually does not occur until Line 30 of the Second Letter. Its proper Text would be 'some works of the Torah' (italics ours). Where the history of Christianity is concerned, this is an important distinction. Our reconstruction, transliteration and Text here are completely new. We have not relied on anyone or any other work, but rather sifted through the entire unpublished corpus, grouping like plates together, identifying all the overlaps, and making all the joins ourselves.

Text

Part 1: Calendrical Exposition

[(1) In the first month, (2) on the fourth (3) of it is a sabbath; (4) on the eleventh (5) of it is a sabbath; (6) on the four- (7) teenth of it is the Passover; (8) on the eight- (9) teenth of it is a sabbath; (10) on the twenty- (11) fifth (12) of it is a sabbath; (13) afterward, on the twenty- (14) sixth (15) of it is the Waving of the Omer. (16) On the second (17) of the second month (18) on that day is a sabbath; (19) on the ninth (20) of it is a sabbath; (21) on the fourteenth (22) of it is the Second Passover; (23) on the sixteenth (24) of it is a sabbath; (25) on the twenty(26) third (27) of it is a sabbath; (28) on the thirtieth (29) of it is a sabbath. (30) In the third month (31) on the seventh (32) of it is a sabbath; (33) on the fourteenth (34) of it is a sabbath; (35) afterward, (36) on the fifteenth (37) of it is the Festival of Weeks; (38) on the twenty- (39) first (40) of it is a sabbath; (41) on the twenty- (42) eighth (43) of it is a sabbath; (44) after (45) Sunday and Monday, (46) an (extra) Tuesday (47) is added. (48) In the fourth month (49) on the fourth (50) of it is a sabbath; (51) on the eleventh (52) of it is a sabbath; (53) on the eight- (54) eenth of it is a sabbath; (55) on the twenty- (56) fifth (57) of it is a sabbath.

(58) On the second (59) of the fifth month (60) is a sabbath; (61) afterward, (62) on the third (63) of it is the Festival of New Wine; (64) on the ninth (65) of it is a sabbath; (66) on] the sixteenth (67) of it is a sabbath; (68) on the twenty(69) third (70) of it is a sabbath; (71) [on the] thir[t]ieth (72) [of it is a sabbath. (73) In the sixth month, (74) on the seventh (75) is a sabbath; (76) on the fourteenth] (77) of it is a sabbath; (78) on the twenty- (79) first (80) of it is a sabbath; (81) on the twenty- (82) second (83) of it is the Festival of (84) (New) Oil; (85) after[ward, on the twenty- (86) third] (87) is the Offerin[g of Wood; (88) on the twenty- (89) eighth (90) of it is a sabbath; (91) after (92) Sunday and Monday (93) an (extra) Tuesday (94) is added. (95) On the first of the seventh (96) month (97) is the Day of Remembrance; (98) on the fourth (99) of it is a sabbath; (100) on the tenth (101) of it is the Day (102) of Atonement; (103) on the eleventh (104) of it is a sabbath; (105) on the fifteenth (106) of it is the Festival (107) of Booths; (108) on the eight- (109) eenth of it is a sabbath; (110) on the twenty- (111) second (112) of it is the Gathering; (113) on the twenty(114) fifth (115) of it is a sabbath. (116) On the second (117) of the eighth month (118) is a sabbath; (119) on the ninth] (120) of it is a sabbath; (121) on the sixteenth (122) of it is a sabbath; (123) on the twenty(124) third (125) of it is a sabbath; (126) on the th[irt]ieth (127) [of it is a sabbath.

(128) In the ninth month, (129) the seventh (130) is a sabbath; (131) on the fourteenth (132) of it is a sabbath; (133) on the tw]ent[y-first (134) of it is a sa]bbath; (135) [on the] twenty- (136) eighth (137) of [it] is a sabbath; (138) after (139) S[unday] and Mond[ay,] (140) [an (extra) Tuesday (141) is added. (142) In the tenth month] (143) on the [fourth (144) of it is a sabbath;] (145) on the eleventh] (146) of' it is a sabbath; (147) on the eight- (148) teenth of it is a sabbath; (149) on the twenty- (150) fifth (151) of it is a sabbath. (152) On the second (153) [of the eleventh] (154) mon[th (155) is a sabbath. (156) on the ninth (157) of it is a sabbath; (158) on the sixteenth (159) of it is a sabbath; (160) on the twenty(161) third (162) of it is a sabbath; (163) on the thirtieth (164) of it is a sabbath. (165-167) In the twelfth month, (168) the seventh (169) is a sabbath; (170) on the fourteenth of it is a sabbath; on the twenty-first of it is a sabbath; on the twenty- (171) eighth of it] is a sabbath; after [Sunday and Monday an (extra) Tuesday (172) is addled. Thus the year is complete: three hundred and [sixty-four] (173) days.

Part 2: Legal Issues

(1) These are some of our words concerning [the Law of Go]d, that is, so[me (2) of the] works that [w]e [reckon (as justifying you; cf. Second Letter, line 34). All] of them have to do with [holy gifts] (3) and purity issues. Now, [concerning the offering of grain by the [Gentiles, who...] (4) and they tou[c]h it... and render it im[pure... One is not to eat] (5) any Gentile grain, nor is it permissible to bring it to the Tem[p]le. [Concerning the sin offering] (6) that is boiled in vessels [of Gentile copper,] by which means [they (the priests) render impure] (7) the flesh of their offerings, and (further, that) they b[oi]l in the courtya[rd of the Temple and thereby

pollute] it (the Temple) (8) with the soup they make-(we disagree with these practices). Concerning sacrifices by Gentiles, [we say that (in reality) they] sacrifice (9) to the i[doll that seduces them; (therefore it is illicit). [Further, regarding the thank] offering (10) that accompanies peace offer[ings,] that they put aside on one day for the next, w[e reckon] (11) that the gra[in offering is to be ea]ten with the fat and the flesh on the day that they are [offer[ed. It is incumbant upon] (12) the priests to assure that care is taken in this matter, so that [the priests will not (13) brin[g] sin upon the people. Also, with regard to the purity of the heifer that purifies from sin (i.e., the Red Heifer): (14) he who slaughters it and he who burns it and he who gathers its ashes and he who sprinkles [the water] (15) (of purification from) sin-all of these are to be pure with the se[tt]ing of the sun, (16) so that (only) the pure man will be sprinkling upon the impure. The sons (17) of Aaron must give wa[rning in this matter...]

(18) [Concerning] the skins of catt[le and sheep...] (19) their [skins] vessels [...One is not (20) to bring] them to the Templ[e...] (21) Also, regarding the skin[s and bones of unclean animals-for they are making] (22) [from the bones] and from the s[kilns handles for ve[ssels-one is not to bring them (i.e., the vessels) to the Temple. With regard to the ski]n from the carcass (23) of a clean [animal,] he who carries that carcass [must not touch [holy items] susceptible to impurity. (24) [...Al]so concerning... that the[y... (30) The members] (31) of the pries[tho]od must [be careful [about] all [these] matters, [so that they will not] (32) bring sin upon the people. [Con]cerning (the fact) that it is written, ['And he shall slaughter it on the side of the altar...,' they (33) are slaughtering] bulls and [lam]bs and shegoats outside the 'camp.' On the contrary, the (lawful) pl[ace of slaughter is at the north within the 'camp.'

(34) We reckon that the Temple [is 'the Tent of Witness,' while] Jerusale[m] (35) is the 'camp.' 'Outside the camp' [means 'outside Jerusalem.'] (It refers to) the 'camp (36) of their cities,' outside the 'ca[mp' (which i]s [Jer]u[salem.) Regarding the si]n offering, they are to remove the offal of (37) [the] altar and bur[n it outside Jerusalem, for] it is the place that (38) [He chose] from among all the tri[bes of Israel, to establish His Name there as a dwelling...] (43)... they are [no]t slaughtering in the Temple. (44) [Regarding pregnant animals, we maintain that one must not slaughter (both)] the mother and the fetus on any one day. (45) [...Also, concerning anyone eating the fetus, w]e maintain that he may eat the fetus (46) [that is in its mother's womb (only) after its (separate) slaughter. You know that th]is is the proper view, since the matter stands written, 'A pregnant animal...'

(47) [With respect to the Ammonite and the Moabite and the bastard and the man with cru[shed testicles and the man with a damaged male organ who are entering (48) the assembly... and taking [wives, to make them 'one (49) bone'... (50) polluted. We also reckon (51) [that one must not... and one must not have inter]course with them (52) [...And one must not integrate them and make them (53) ['one bone'... And one must not bring] them in (54) [...And you know that so]me of the people (55) [...integr]ating (56) [...For the sons of Israel must guard

against] all illicit marriage (57) and (thus) properly revere the Temple. [In addition, concerning the bli[n]d, (58) who cannot see so as to avoid polluting mingl[ing,] and to whom [sinful (59) mingling is invisible-(60) as well as the deaf, who hear neither law, nor statute, nor purity regulation, and do not (61) hear the statutes of Israel-for 'He who cannot see and cannot hear does not (62) know how to perform (the Law)'-these people are trespassing on the purity of the Temple!

(63) Concerning poured liquids, we say that they possess no (64) intrinsic [pu]rity. Poured liquids do not (properly) separate between the impure (65) and the pure (i.e. vessels), because the fluid of poured liquids and that of a receptacle used with them (66) is one and the same (i.e., the pollution travels between the vessels along the path of the fluid). One is not to bring dogs into the H[ol]y 'camp' because they (67) eat some of the bones in the Te[m]ple while the flesh is (still) on them. Because (68) Jerusalem is the Holy 'camp'-the place (69) that He chose from among all the tribes of Israel. Thus Jerusalem is the foremost of (70) the 'ca[m]ps of Israel.' Regarding trees planted for food (71) in the land of Israel, (the fruit of the fourth year) is analogous to a first fruit offering and belongs to the priests. Likewise the tithe of cattle (72) and sheep belongs to the priests. In the matter of those suffering from a skin disease, we (73) s[ay that they should not dome with holy items susceptible to impurity. Rather, they (74) must stay alone [outside the camps. And] it is also written, 'From the time when he shaves and bathes, let him [st]ay outside (75) [his tent seven d]ays.' But at present, while they are still impure, (76) those suffering from a skin disease are coming] home [w]ith holy items susceptible to impurity. You know (77) [that anyone who sins by inadvertence, who breaks a commandment] and is forgiven for it, must bring (78) a sin offering (but they are not doing so). [As for the intentionally disobedient, it is written, 'He is a despiser and a blasphemer.'

(79) [While th]e[y suffer impurities caused by [s]kin diseases, they are not to be fed with ho[ly] food (80) until the sun rises on the eighth day (after they are cured). Concerning [impurity caused by contact with a dead] (81) person, we say that every (human) bone, whether it is [skeleton] (82) or still covered (with flesh), is governed by the statute for the dead person or those slain in battle. (83) As for the fornication taking place among the people, they are (supposed to be) a (84) Holy People, as it is written, 'Israel is Holy' (therefore, it is forbidden). Concerning a man's cloth[es, it is written, 'They are not] (85) to be of mixed fabric;' and no one should plant his field or [his vineyard with mixed crop]s.

(86) (Mixing is forbidden) because (the people) is Holy, and the sons of Aaron are H[oly of Holy](87) [nevertheless, as y]ou know, some of the priests and the [people are mixing (intermarrying).] (88) [They] are intermarrying and (thereby) polluting the [holly seed, [as well as] (89) their own [see]d, with fornication...

The Second Letter on Works Reckoned As Righteousness (4Q397-399)

That this text is a second letter is clearly signalled in Lines 29-30, quoted above, which refer to a first letter already having been written on the same subject - 'works reckoned as justifying you' (italics ours). Though fragments of the two letters are in the same handwriting, it is not clear that these are directly connected or on the same or succeeding columns.

The text ends with a ringing affirmation of what can be described as the Jamesian position on 'justification': that by 'doing' these 'works of the Law' however minute (note the emphasis on doing again) in the words of Gen. 15:6 and Ps. 106:31 - a psalm packed with the vocabulary we are considering here - 'it will be reckoned to you as Righteousness'. As a result, you will have kept far from 'the counsel of Belial' and 'at the End Time you will rejoice' (32-3). This last most surely means either 'being resurrected' or 'enjoy the Heavenly Kingdom', or both - an interesting proposition to be putting to a king or Community Leader in this time.

Text

(2)... because they come... (3) will be... (4) and concerning wome[n...] And the rebellion [...(5) For by reason of these... because of] violence and fornication [some] (6) places have been destroyed. [Further,] it is writt[en in the Book of Moses,] 'You [are no]t to bring the abomination t[o your house, because] (7) the abomination is despised (by *God*).' [Now, you know that] we broke with the majority of the peo[ple and refused] (8) to mix or go along wi[th them] on these matters. You also k[now that] (9) no rebellion or Lying or Evil [should be] found in His Temple. It is because of [these things w]e present [these words] (10) [and (earlier) wrot]e to you, so that you will understand the Book of Moses [and the words of the Prophets and of Davi[d, along with the (11) chronicles of every] generation. In the Book (of Moses) it is written .... s[o] that not...

(12) It is also written, '[(If) you turn] from the W[a]y, then Evil will meet [you.'] Again, it is written, (13) 'It shall come to pass that when [al]l [t]hese thing[s com]e upon you in the End of Days, the blessing (14) [and] the curse [that I have set before you, and you call them to m[in]d, and return to me with all your heart (15) and with [a]ll [your] soul' [...at the En]d [Time,] then you will l[i]v[e... Once again, (16) it is written in the Book] of Moses and in [the words of the Prophets that [blessings and curses] will come [upon you... (21) the ble]ssin[gs that] cam[e upon i]t (Israel) in [his days [and] in the days of Solomon the son of David, as well as the curses (22) [that] came upon it from the d[ays of Jer]oboam the son of Nebat until the exi[l]e of Jerusalem and Zedekiah the king of Jud[ah.]

(23) [For] he may bri[n]g them upon... And we recognize that some of the blessings and curses have come, (24) those written in the Bo[ok of Moses; therefore this is the End of Days, when (those) in Isra[e]l are to return (25) to the La[w of *God* with

all their heart,] never to turn bac[k] (again). Meanwhile, the wicked will increase in wick[ed]ness and...

(26) Remember the kings of Israe[l], and understand their works. Whoever of them (27) feared [the L]aw was saved from sufferings; when they so[ug]ht the Law, (28) [then] their sins [were forgiven] them. Remember David. He was a man of Pious works, and he, also, (29) was [salved from many sufferings and forgiven. And finally, we (earlier) wrote you about (30) some of the works of the Law (see the First Letter above), which we reckoned for your own Good and for that of your people, for we see (31) that you possess discernment and Knowledge of the Torah. Consider all these things, and beseech Him to grant you (32) proper counsel, and to keep you far from evil thoughts and the counsel of Belial.

(33) Then you will rejoice at the End Time, when you find some of our words were true. Thus, 'It will be reckoned to you as Righteousness' (or in Paul's language, 'reckoned as justifying you'), your having done what is Upright and Good before Him, for your own Good and for that of Israel.

A Pleasing Fragrance (Halakhah A - 4Q251)

This text is typical of the kind of legal minutiae found at Qumran. It further fleshes out our view of the basic legal approach there. In it, there are parallels to both the Community Rule and the Damascus Document. For instance, the enumerations contained in Fragment 1 parallel many in the Community Rule.

Those in Fragment 2 parallel similar materials in the Damascus Document. In both cases the parallels are precise, though the language varies. For instance, the penalty for 'knowingly Lying' in Line 7 of Fragment 1 and in the Community Rule are exactly the same, though the offence is described slightly differently (1QS, vii. 3).

Even more interesting in this text are the descriptions of the Council of the Community. The language of 'making atonement for the land' and the reconstructed material about being 'founded on Truth for an Eternal Planting' in 3.8-9 is exactly that of the Community Rule, viii.5-6. If the reconstruction 'fifteen men' in Line 3.7 is correct, then another puzzling problem in Qumran studies is solved - whether the 'twelve men and three priests' mentioned in the Community Rule, viii. 1 as 'a Holy of Holies' and 'a House (i.e. a Temple) for Israel' should be exclusive or inclusive: that is, should they be three within the twelve, paralleling Christian reckonings of twelve apostles and a central triad, or three in addition to twelve?

Text

Fragment 1

(1) [...t]en days... (2) thirty days [...he will be fined] (3) half of his food ration for fift[een days...] (4) he will be fined for three months [half his food ration. A man that speaks prior to the turn] (5) of his neighbor, though he (the latter) is enrolled ahead of him, must be separated [from the Many...] (6) in them half his food ration. A man who... (7) thirty days. And a man who tells a lie kn[owingly will be punished for six] (8) months, and fined during that time half his food ration... (9) knowingly about everything, his penalty is thirty days and half his food ration... (10) kno[wingly, they shall] separate him (from the Community) for six months...

Fragment 2

(1) The Sab[bath...] (2) on the Sabbath d[ay]. A [man] should not [wear garments that are] soil[ed on the Sabbath day]. (3) A man should not... in garments th[at have] dust on them or... (4) on the Sabbath day. [A ma]n [should not bring out] of his tent any vessel or food (5) on the Sabbath day. A man should not lift out cattle which has fallen (6) in[to] the water on the Sabbath day. But if it is a human being who has fallen into the wat[er (7) on the day of] the Sabbath, he will throw him his garment to lift him out with it. But he will not lift an implement... (8) [...on the day of] the Sabbath. And if an army...

Fragment 3

(1)... on the day of [the Sabbath...] (2) on the day of] the Sabbath, and not... (3) A man from the seed of Aaron will [no]t sprinkle the waters [of impurity on the Sabbath day...] (4) [The Passover is a (high) holiday, and a fast on the day of [the Sabbath .... A man] (5) may take his cattle two thousand cub[its on the Sabbath day, but he may not walk unless he is a distance from] (6) [the Temple of more than thirty stadia.. A [man] should not... (7) [When] there are in the Council of the Community fif[teen men, Perfect in everything which has been revealed in all the Law] (8) [and the Prophets, the Council of the Communi[ty] shall be founded [on Truth for an Eternal Planting and true witnesses at the judgement, and the Elect] (9) of (*God*'s) favor, and a pleasing fragrance to make atonement for the land, from a[ll Evil...] (10) The Period of Wickedness will end in judgement, and the... (11) In the first] week... (12) which were not to brought to the Garden of Eden. And the bone of... (13) shall be for it forever, which was not brought nea[r... (14) to the] Holiness of the Garden of Eden, and all the verdure in its midst is Holy. [When a woman conceives and bears a boy,] (15) she shall be unclean seven days. Just as in the days of her menstrual impurity, she shall be unclean; and th[irty three days she shall remain in the blood of (16) her purification. If she bears a girl, she will be unclean [two weeks, just as in her menstrual period; and sixty-six days (17) she shall rema]in in the blood of her purification. [She should touch] no Holy thing [and should not enter the Temple.]

Fragment 4

(1) [...If a man strikes another] in the eye, [and he is bedridden, but gets up and walks around (2) outside on his staff, the one who struck him will merely] compensate [him for his] con[valescence and his medi]cal expenses. (3) [If a bull gores a man or wo]man, the bull will be killed. He will stone it, (4) [and not eat its meat, and the bull's owner shall be blameless. But if the bull had gored previously (5) [and the owner had been apprised, but he had not kept it (penned up); and it kills a m]an or a woman, (6) [the bull will be stoned, and its owner will be put to death as well... .]

Fragment 5

(1) [...grain, new wine, or fresh oil, unless [the priest has waved it...] (2) their early produce, the frst fruits. And a man should not delay (giving) the full measure, for... (3) is the firstfruit of the full measure. [And the] grain is the offering... [And the bread of] (4) the firstfruits is the leavened bread that they bring [on] the day of the [firstfruits...] (5) are the firstfruits. A man should not eat the new wheat... (6) until the day he brings the bread of the firstfruits. He should not...

Fragment 6

(2)... he should not... (3) [...the grain offering of the tithe is for [the priest...] (4) [...the firstborn of a ma]n or uncle[an] cattle [he should redeem] (5) [...the firstborn of a man or unclean cattle (6)... the flock, and the sanctuary, from (7) [...i]t is like the firstborn, and the produce of a tree (8) [...in the first [year], and the olive tree in the fourth year (9) [...and the heave offering; everything that is set apart for (the support of) the priesthood

Fragment 7

(1) Concerning immoral unions... (2) A man should not marry his si[ster, the daughter of his father or the daughter of his mother... A man should not marry] (3) the daughter of his brother or the daughter of his si[ster...He should not uncover] (4) the nakedness of the sister of his fa[ther or the sister of his mother. Nor should a woman be given to the brother of] (5) her father or to the brother of her mother [to be his wife...] (6) A man should not uncover the nakedness of... (7) A man should not take the daughter of...

Mourning, Seminal Emissions, Etc. (Purity Laws Type A - 4Q274)

The contents of this manuscript are remarkable, discussing issues not previously found in any Qumran text. All have to do with matters of ceremonial purity and impurity in accordance with stipulations set forth in legal portions of the Bible. A number of the matters considered here involve impurities that require a seven-day period of purification.

Provisions are made for mourners to dwell apart in a special place during the days of their impurity, apparently acquired from contact with the corpse of a loved one. The text stipulates that such people and others, suffering from Levitical uncleanness of various sorts, are to dwell to the north-west of the nearest habitations - a law reminiscent of the placement of the latrine in the Temple Scroll (11QT, xlvi.13-16).

Text

Fragment 1

Column 1

(1) he is to delay distributing the portions (that he has prepared for the priests). He is to sleep in a bed of mourning and dwell in the house of bereavement, separated with all the other impure persons, twelve cubits distance from (2) the pure food, in the designated part of town, and the same distance to the northwest of any inhabited dwelling. (3) Any man suffering from the various types of impurity should bathe himself and wash his clothing on the [seve]nth [da]y, and afterwards he may eat (the pure food). For this is what it means, "Unclean, unclean!' (4) he should call all the days of his affliction.' As for the woman who suffers a seven-day flux of blood, she should not touch a man suffering from a flux, nor any implement that he touches, (5) nor anything upon which he rests. But if she does touch (these things), she should wash her clothes and bathe and afterwards she may eat (the pure food). At all costs she is [no]t to mingle during her seven (6) days, so that she does n[o]t defile the camp of the Ho[ly O]nes of Israel. Nor is she to touch any woman suffer[ing] a long[standing] flux. (7) And the person that is keeping a record of the period of impurity, whether a man or a woman, is not to to[uch the menstruant] or the mourner during the period of uncleanness, but only when she is cleansed [from her uncleanness, for (8) that uncleanness should be reckoned in the same way as a flux [for] anyone who touches it. And if someone touches a [bodily] flux [or a seminal emission, then [h]e should be unclean. Anyone touching any of these types of (9) impure people should n[ot] eat the pure food during the seven days of his purification]. When someone is impure because of touching a core[se, he is to bathe himself in water and wash (his clothes) and afterwa[rds]

Column 2

(1) he may a[at...] (2) and sem[inal emission...]

Fragment 2

Column 1

(1)... who sprinkle upon him for the first time and he should bathe and wash his clothes before (2)... on the seventh day. One is not to sprinkle on the sabbath (3)... on the sabbath, only he should not touch the pure food until he changes (4) [his clothes...] Anyone who touches a human seminal emission must immerse everything down to the last item of dress, and the person that carries the item (5) [...must immerse, and the garment upon which the emission is found or any item that carries it (the emission) (6)... And if there should be found in the camp any man that is incapab[le] (or, that does not have enough) (7)... the garment that it/she has not touched, only he should not touch it-his meat-and he that touches (8)... they should dwell. If he did not touch it, he should wash it in water, but if (9) [he did touch it]... and he should wash (it). Regarding all offerings, a man should wash

Column 2

(1) his flesh, and thus... (2) but if... (3) with him... (4) to... (5) reptile. Impure [people...] (6) and he that touches it... (7) and eve[ry...] (8) But if... (9) who...

Fragment 3

Column 1

(1)... *God's* revealing the apple of His eye, and (2)... every law... (3) or every... (5) and she is unclean... (6) they pour liquid upon and he does not eat eat in purity, and every... (7) [(everything) tha]t they will dissolve by rubbing, and whose solvent liquid has evaporated, a man should not eat (8)... the impure among them, and also among garden vegetables... (9) or a boiled cucumber. A man who [po]urs liquid upon a foodstuff

Column 2

(1)...they are impure... (3) and everything that he possesses... (4) to purify, and the remains of all the garden vegetables (5) from the moisture of dew he may eat, but if n[ot]... (6) in the water, except a man... (7) the land, if there comes upon it... (8) the rain upon it, if a man touches it... (9) in a field by all means he at the turn of the season of... (10) every frangible vessel that... (11) which is in its middle... (12) the foodstuff upon which water has been poured...

Laws of the Red Heifer (Purity Laws Type B - 4Q276-277)

The two manuscripts presented here are of the most interesting subject matter: the law of the Red Heifer, a subject already referred to in the First Letter on Works Reckoned as Righteousness above. The Red Heifer purification ceremony was one of the most holy in Jewish tradition.

According to Numbers, the people were to prepare this heifer by burning it with cedar wood, hyssop and scarlet material. The ashes were then gathered and mixed with water. This water, known as 'the water that removes impurity', was sprinkled upon those who had acquired certain types of ritual impurity - including some of the bodily and sexual impurities mentioned in the preceding text.

Text

Manuscript A

Fragment 1

(1)[...clothes] in which he has not performed (any) sacr[ed] rite. (2)... and he should regard the clothes as impure(?). Then he (a designated man, not the presiding priest) will slaugh[ter] (3) [the] heifer before him, and he (the priest) will take up its blood in a new earthen vessel that (4) [has never drawn] near the altar and sprinkle some of its blood with [his] finger seven (5) [times to]ward the front of the T[e]nt of Meeting. Then he should cast the cedar wood (6) [and the hyssop and the scarlet] material into the mid[st] of its (the heifer's) fire. (7) [Then the priest and the man who burns (the heifer) and the man who gathers the heifer's ashes [should bathe] (8) [and wash their clothes, and they shall be unclean until evening. And this they should establish as a ceremony (9) [for that water which removes the impurity of sin, and as an Eternal Law. And] the priest should put on

Manuscript B

Fragment 1

(1)... and the hyssop and the... (2) [a man] pure of all sin[full impurity... (3) [And] the priest who atones with the blood of the heifer and all [the men should] put [on different clothes (4) and wash their tu]ni[cs and] their seamed robes in which they made atonement in performing the law [governing the removal of sin. (5) Each man should bathe] in wader and be un]clean until the eve[n]ing. The man who carri[es the plot containing the water that removes impurity will be im[pure...] (6) A man [should sprinkle] the water that removes impurity upon those who are imp[ure, in]deed a pu[re] priest [should sprinkle (7) the water that removes impurity on the]m. Thu[s he will] atone for the impure. No wicked man is to sprinkle upon the impure. A m[an] (8) [...the water that removes im]purity. They should enter the water and become pu[r]e from the impurity that comes from contact with dead people... (9) [an]other. [The priest [should] sprinkle them with the water that removes impurity, to purify (10) [...R]ather they shall become pure, and their [fl]esh p[ure.] Anyone who touches [him...] (11) his flux... and their [hands] unwashed with water, (12) [then] they shall be impure... his be[d] and [his] sea[t...] they touched his [f]lux, (it) is like the impurity [that comes from contact with dead people.] (13) [The] man who touches (these things) [should bathe and be] impure till [the] evening, and the man who carries (these things) [should wash] his [cl]othes and be impure till evening.

The Foundations of Righteousness (The End Of The Damascus Document: An Excommunication Text - 4Q266)

There can be no doubt that what we have here is the last column of the Damascus Document. Though the text as we have it here does not precisely follow any material from either of the two overlapping known manuscripts found by Solomon Schechter in the Cairo Genizah in 1897, many of its allusions do. So does their spirit.

The piece is preserved in two copies: one is nicely ruled; the second, in what is called 'semi-cursive', would appear to be a private copy. We present the second, which is the more completely preserved, with occasional help from the first to fill in blanks. That it really is the last column of the document is ascertainable from the blank spaces on the parchment at the left. The edge of a previous column with some stitching is also visible on the right (see Plates 19-20).

This language of 'doing the exact sense of the Torah' is very important. It is also to be found earlier still in vi. 14-15 coupled with a reference to 'the Era of Evil' and 'separating from the sons of the Pit' (italics ours). This 'not one jot or tittle' approach to the Law is, of course, prominent in traditions associated with Jamesian Christianity, not least of which is the famous condemnation of 'breaking one small point of the Law' in James. 2:10. Here, too, 'doing' and 'breaking the Law' are prominently mentioned. The text ends by evoking the phrase midrash ha-Torah, i.e. 'the study' or 'interpretation of the Law'.

Text

[...before the Priest commanding]

(1) the Many, and he freely accepted His judgement when He said by the hand (2) of Moses regar[ding] the person who sins inadvertently, 'let such a one bring (3) [his] sin offering [or] his guilt offering.' And concerning Israel it is written, 'I shall ascend (4) to [the Highest in Heaven, and there will not smell the fragrance of their

offerings.' And in another place (5) it is written, 'return to *God* with weeping and fasting.'

(5ª) (In [anoth]er place it is written, 'rend your hearts and not your clothes.') As for every person who rejects these (6) Judgements (which are) in keeping with all the Laws found in the Torah of Moses, he will not be reckoned (7) among all the sons of His Truth, for his soul has rejected the Foundations of Righteousness. For rebellion, let him be expelled (8) from the presence of the Many. The Priest commanding the Many shall speak against him. He (the Priest) is to stand (9) and say, 'Blessed are You, You are all, everything is in Your hand and (You are) the maker of everything, who established (10) [the Peoples according to their families and their national languages. You 'made them to wander astray in a wilderness without a Way,' (11) but You chose our fathers and to their seed gave the Laws of Your Truth (12) and the judgements of Your Holiness, 'which man shall do and thereby live.' And 'boundary markers were laid down for us.' (13) Those who cross over them, You curse. We, (however), are Your redeemed and 'the sheep of Your pasture.'

(14) You curse their transgressors while we uphold (the Law).' Then he who was expelled must leave, and whosoever (15) eats with him or asks after the welfare of the man who was excommunicated or keeps company with him, (16) that fact should be recorded by the Mebakker / Overseer according to established practice and his judgement will be completed. The sons of Levi and (17) [the inhabitants] of the camps are to gather together in the third month (every year) to curse those who depart to the right or (18) [to the left from the] Torah. And this is the exact sense of the judgements that they are to do for the entire Era (19) [of Evil, that which was commanded [for al]l the periods of Wrath and their journeys, for everyone [who dwells in their camps and all who dwell in their cities, al]l that [is found in the 'Final M]idra[sh] of the Law.'

Fernando Klein

Commentaries

Isaiah Commentary (4QIsaiah Pesher, 4Q161, 4QpIsa)

Pesher is a kind of commentary on the Bible that was common in the community that wrote the Dead Sea Scrolls. This kind of commentary is not an attempt to explain what the Bible meant when it was originally written, but rather what it means in the day and age of the commentator, particularly for his own community. In the Isaiah Pesher, or commentary on the book of Isaiah, a verse or verses from Isaiah are quoted. Then the commentary begins, often introduced by the word "pesher," or "the interpretation of the word..." If we were to write a commentary in this way today we might quote a bible verse and then say, "and the meaning of the verse is..." and go on to show the significance of the verse for our own church, synagogue, or society.

This particular manuscript quotes several verses from Isaiah 5 concerning punishment or destruction, and applies them to the "arrogant men" who are in Jerusalem. We know from other scrolls at Qumran that the people who wrote many of the scrolls had serious conflicts and disagreements with the religious leaders in Jerusalem over the proper way to conduct worship in the Temple. Most scholars think that the community of the Dead Sea Scrolls was led by a group of priests who thought that the Jerusalem priests were corrupt. The group at Qumran therefore started their own community in which they tried to live pure and righteous lives, away from the corrupting influence of Jerusalem.

4QIsaiah Pesher, 4Q161, 4QpIsa

Frag. 1 col. I 20 Is 10:20 [On that day, the remnant of Israel, the survivors of Jacob, will not revert to leaning] 21 [on their assailant but will lean exclusively on the Lord, the Holy One of Israel.] 22 [Is 10:21 A remnant will return, a remnant of] Jacob to God [the warrior.] 23 [Its interpretation: the remnant of] Israel is [the assembly of his chosen one...] 24 [...] the men of his army [...The remnant of Jacob is...] 25 [...] the priests, since [...]

Frags. 2-6 col. II 1 [Is 10:22 Even if your people, Israel were like the sand of the sea, only a remnant will return; extermination is decreed,] 2 [but overflowing justice. For it is decided and decreed: the Lord of Hosts is going to do it in the centre of all the earth.] 3 [Its interpretation concerns...] since [...] the sons of [...] 4 of his people. And as for what he says: Is 10:22 Even if [your people, Israel were like the sand of the sea,] 5 [only a remnant will return; extermination is decreed,] but justice will overflow. [Its interpretation:...] 6 [...to des]troy on the day of slaughter]; and many will die [...] 7 [...but they will be] saved, surely, by their plan[ting] in the land [...] 8 [...] Blank Is 10:24-27 This is why [the Lord God of Hosts] says: [do not be afraid, my people] 9 [li]ving in Zion, [of Assyria: it will hit you with a stick and lift its rod against you in the fashion of Egypt;] 10 [for] very shortly [my anger will end and my wrath will destroy them. The Lord of Hosts] will lash [against them] 11 [the flail as in the destruction of Midian, on the rock of]

Horeb, and he will lift his rod [against the sea] 12 [in the fashion of Egypt. And on that day it will happen] that his load will be removed [from your shoulder,] 13 [and his yoke from your neck. The interpretation of the word concerns] ... [...] 14 on his return from the wilderness of the [peoples...] 15 [...] the Prince of the Congregation, and after it will be removed from you [...] 16 [...] Blank [...] 17 [Is 10:28-32 Go up from the side of Rimmon;] come up to Aiath; cross [Migron; at Michmash] 18 [make an inspection of the weapons; traverse] the gorge; spend the night in Geba; fearful [is Ramah; Gibeah of] 19 [Saul deserts. Raise] your voice, Bat-Gallim; pay attention, Laishah; answer, Anathoth.] 20 [Retreat,} Madmenah, the residents of Gebiom flee; this very day [he makes a stopover in Nob,] 21 [already he stretches] his hand towards the mount of the daughter of Zion, towards the hill of Jerusalem. [Blank] 22 [The interpretation of] the word concerns the final days, when the [king of the Kittim] comes [...] 23 [...] from his climb from the plain of Akko to do battle against Pa[lestine...] 24 [...] and there is none like her, and in all the cities [...] 25 and up to the boundary of Jerusalem. [...]

Frags. 8-10 col. III 1 [Is 10:33-34 See! The Lord God of Hosts will rip off the branches at one wrench; the[tall[est trunks] will be felled, 2 [the loftiest chopped.] The thickest of the wood [will be cut] with iron and Lebanon, with its grandeur, 3 [will fall. Its interpretation concerns the] Kittim, who will be placed in the hands of Israel, and the meek 4 [of the earth...] all the peoples and all the soldiers will weaken and their heart will melt 5 [...and what it says: { The] tallest [trunks] will be destroyed} are the soldiers of the Ki[ttim] 6 [since ...] {and the thickest of the wood will be cut with iron} are 7 [...] for the war of the Kittim. {And Lebanon, with its grandeur, 8 [will fall} are the commanders of} the Kittim, who will be placed in the hand of their great [...] 9 [...] in their flight before Israel. 10 [...] Blank [...] 11 [Is 11:1-5 A shoot will issue from the stu]mp of Jesse and [a bud] will sprout from its ro[ot.] Over him [will be placed] the spi[rit] 12 [of the Lord; a spirit] of discretion and wisdom, a spirit of ad[vice and courage,] a spirit of knowledge 13 [and of respect for the Lord, and his delight will be in respecting the[Lord. [He will not judge] by appearances 14 [or give verdicts on hearsay alone;] he will judge [the poor with justice and decide] 15 [with honesty for the humble of the earth. He will destroy the land with the rod of his mouth and with the breath of his lips] 16 [he will execute the evil. Justice will be the belt of] his loins and lo[yalty the belt of his hips.] 17 [...] Blank [...] 18 [The interpretation of the word concerns the shoot] of David which will sprout [in the final days, since] 19 [with the breath of his lips he will execute] his enemies and God will support him with [the spirit of] courage [...] 20 [...] throne of glory, [holy] crown and hemmed vestments 21 [...] in his hand. He will rule over all the people and Magog 22 [...] his sword will judge all the peoples. And as for what he says: {He will not 23 [judge by appearances] or give verdicts on hearsay}, its interpretation: 24 [...] according to what they teach him, he will judge, and upon his mouth 25 [...] with him will go out one of the priests of renown, holding clothes in his hand

4QIsaiah Pesher (4Q162 [4QpIsb])

Frag. 1 col. I Is 5:5 For now I will tell you what I am going to do with my vineyard: 1 [...remove its fence so that it can be used for pasture, destroy] its wall so that you trample it. Is 5:6 For 2 [I will leave it flattened; they shall not prune it or weed it, brambles and thi[stles] will grow. the interpretation of the word: that he has deserted them 3 [...] and as for what he says: Is 5:6 {Brambles will grow, 4 [and thistles]}: its interpretation concerns...] and what 5 [it says:...] of the path 6 [...] his eyes

Frag. 1 col. II 1 The interpretation of the word concerns the last days, laying waste the land through thirst and hunger. This will happen 2 at the time of the visit to the land. Is 5:11-14 Woe to those who rise early in search of intoxicants and carry on until by twilight the wine 3 excites them and with zithers, harps, tambourines and flutes they feast their drunkenness, but they pay no attention to God's doings 4 or notice the works of his hands! For this, my people will be exiled without realising it, their nobles will die of hunger 5 and the ordinary folk have a raging thirst. For this, the abyss distends its jaws and enlarges its mouth immeasurably, 6 lowers its nobility and its ordinary people and its revelling throng enters. These are the arrogant men 7 who are in Jerusalem. They are the ones who: Is 5:24 {Have rejected the law of God and mocked the word of the Holy One of 8 Israel. Is 5:25 For this the wrath of God has been kindled against his people and he has stretched out his hand against them and wounded the. 9 The mountains quake, their corpses lie like dung in the middle of the streets. In spite of this 10 [his anger] is not appeased [and his hand continues to be stretched out]}. This is the Congregation of the arrogant men who are in Jerusalem. 11 [...] ... [...]

Frag. 1 col. III 1 Is 5:29-30 and no-one rips [it out. On that day he will roar against him] 2 like the ro[ar of the sea. He will look at the earth, see deep darkness, even the light is obscured] 3 by the clo[uds...] 4 He is [...] 5 they [...] 6 who co[me...] 7 he has said [...] 8 they have seen [...] 9 ... [...]

4QIsaiah Pesher (4Q163 [4QpIsc])

Frag. 1 1 [...] ... [...] 2 [...] he is [...] 3 [...] and mistook the path of [...] 4 [...for it is] written concerning him in Jere[miah...]

Frags. 2-3 1 [Is 8:7-8 For this, behold, the Lord will bri[ng up against them the [torrential and violent] water of the river, [the king of Assyria] 2 [and all his pomp. He will] come up through all the channels and overflow all its banks. [He will invade Judah, he will flood, he will brim over] 3 [and will reach right up to the neck.] The opening of his wings will cover the breadth of your land, [O Emmanuel! The interpretation of the word concerns] 4 [...] ...the law; he is Rezin and the son of [Romeliah...] 5 [...as it is written in [...] 6 [...] and not [...]

Frags. 4-6 col. I 14 [Is 9:17-20 Because evil is burning like a fire] which consumes thistles [and brambles;] it catches fire 15 [in the dense wood and the height] of the smoke coils upwards. [By the wrath of the God] of Hosts devastated is] 16 [the land and the people is fuel for the fire.] No-one [forgives] his brother,] 17 [he destroys to the right and remains hungry, he consumes] to the left and is not replete; 18 [a man eats the flesh of his arm. Manasseh against] Ephraim and Ephraim against 19 [Mana]sseh; [the two] together [against Judah. And with all this] his wrath is not mollified.

Frags. 4-6 col. II 1 [Is 10:19 A young man] will count them. [...] 2 The interpretation of the word concerns the edict of Babylonia [...] 3 the edicts of the peoples [...] 4 to betray many. He [...] 5 Israel. And what it says: Is 10:19 {The remainder of the trees of the wood will be a small number and a young man will count them}.] 6 Its interpretation concerns the reduction of men [...] 7 Blank [...] 8 Is 10:20-22 On that day it will happen [that the remainder of the House of Israel and the survivors] 9 of the House of Jacob [will not return to lean on their aggressor but will lean on the Lord, the Holy One] 10 of Israel, in truth. A remnant [will return, a remnant of Jacob, to the warrior God.] 11 Even is your people, [Israel] were to be [like the sand of the sea, only a remnant will return.] 12 The interpretation of the word concerns the fi[nal] days [...] 13 they will go into captivity [...And what] 14 it says: Is 10:21 [{Even if your people, [Israel] were to be like the sand of the sea, only a remnant will return.}] 15 Its interpretation concerns the reduction [...] 16 Since it is written: [Is 10:22-23 {Destruction is decreed but justice is overflowing. Because destruction is decreed] 17 the Lord God [of the Hosts will execute it in the midst of the whole earth}] 18 Blank [...] 19 is 10:24 Therefore, the Lord Go[d of Hosts says as follows: do not fear, my people who live in Zion]

Frags. 8-10 1 [The interpretation of the word] concerns the king of Babylon, [since...as it is written: Is 14:8 {The very cypresses] 2 [laugh] at you, and the cedars of Lebanon. Since [you lie down, the hewer] 3 [does not come up] against them}. The cypresses and the cedars [of Lebanon are...] 4 [...] the Lebanon. And what it says: Is 14:26-27 {This [is the strategy decided for] 5 [all] the earth and this is the hand [stretched out against all the peoples.] 6 [For the God] of the Hosts has dec[ided, who will thwart him? His hand is stretched out,] 7 [who] will push it aside?}. This is... [...] 8 [as it is written] in the book of Zechariah... [Zac 3:9?...] 9 [...] Blank [...] 10 [...] Blank [...] 11 [Is 14:28-30) In the year of the deat]h of king Achaz [this oracle was uttered: Do not] rejoice, 12 al[l Pilistia,] that the rod [which injured you] is shattered, [because from the root of the] snake shall [come] 13 [a viper and its fruit will be a] flying [asp. The most destitute] will be fed [and the poor] 14 [will become safe. I will make your root die of hunger and he will kill] your remnant. [...]

Frag. 11 col. I 1 [...] ... 2 [...] servants of 3 [...] they are 4 [...] the insults (?) 5 [...] this Frag. 11 col. II 1 Is 19:9-12 those who weave [white cloths. Their masters will be dismayed, all their] 2 labourers knocked [down. How deranged the princes of Zoan; the wise advise] 3 Pharaoh with [inane] advice. [How can you say to

Pharaoh: We are sons of wise men,] 4 we are sons of [ancient ki]ngs? [Where are your wise men? Let them announce,] 5 [if they know, what the God of Hosts is planning against Egypt.]

Frags. 15-16 Is 29:10-11 [For] 1 the Lord pours [upon] you [a breath] of languor and will blinker [your eyes-the prophets-and] 2 he will cover your heads-the seers--. For you [any vision] will be [like the text of a] 3 [sea]led [book:] they give it to someone who can read, telling[g him: Please read this,] 4 [and he answers: I cannot because] it is sealed. [...]

Frags. 18-19 1 Is 29:18-23 [of the book;] without darkness or glo[om the eyes of the blind will see. The oppressed will return to rejoice in the Lord] 2 [and the poor]est of men [will delight in the Holy One of Israel. Because the tyrant is destroyed, the sceptic finished off and] 3 [all] those alert for evil [will be obliterated, those who are going to seize another in speaking and the one who defends in the gate with snares and, for nothing, engulf] 4 [the innocent.] Therefore, so says [the Lord to the House of Jacob, he who ransomed Abraham: No longer] 5 [will] Jacob [be ashamed, no longer will his face smile when he sees that his sons, the work] 6 [of my hands in his midst,] worship my name, because they wor[ship the Holy One of Jacob.] Frag. 21 1 [...] Is 29:17 Perhaps, [in a very little while, 2 will the Lebanon turn into] and orchard, and will the orchard] seem like [a wood?] The Lebanon are [...] 3 [...] into an orchard and they will turn into [...] 4 [...] by the sword. And what it [says...] 5 [...] ... [...] 6 [...] ... [...] the teacher of [...as it is written:] 7 Zach 11:11 [It was annulled on that day, and[thus the most helpless of the flock which [was watching me knew] 8 [that it was in fact the word of the Lord.] Blank [...] 9 Is 30:1-5 [Woe to the rebellious sons-oracle of the] Lord-who make plans [without counting] 10 [on me; who sign deals, but[without my spirit, to ad[d sin] 11 [to sin; who proceed to go do]wn to Egypt [without conferring with me, to gain strength] 12 [with the strength of the Pharaoh and shel]ter in the shadow of Egypt! [Their disgrace will be] 13 [the strength of the Pharaoh, and the she]lter of the shadow of Egypt, [their shame. For in Zoan were] 14 [their princes, and their messengers] reached Hanes. [They were all ashamed of a] 15 [powerless people which could neither help] nor [oblige...]

Frag. 23 1 [...] and they [...] all [...] ... [...] 2 [...] ... [...] Blank [...] 3 Is 30:15-18 [For] thus says YHWH, the Holy One of Israel: By turning back and being pla[cid will you be save;] 4 your courage will comprise composure and trust. But you did not wish and sa[id:] 5 No, let us flee on horseback. Well, then, you need to flee. We will run at a gallop. Well, then 6 those chasing you will run faster. A thousand [shall flee] before the menace of one, before the menace 7 of five shall you flee, until you end up like a flagpole on the peak of a mountain, 8 like a standard upon a hill. This is why the Lord waits to take pity on you, this is why he rises 9 to be lenient with you. For YHWH is a God of justice. Happy are those waiting for! 10 The interpretation of the word, for the last days, concerns the congregation of those [looking] for easy interpretations 11 who are in Jerusalem [...] 12 in the law and not [...] 13 the heart, for in order to crush [...] 14 Hos 6:9 As bandits lie in wait, [the priests scheme]. They have rejected the law [...] 15 Is

30:19-21 [F]or a people living in Zion, [in Jerusalem, will no longer need to weep; the voice will have pity on you at the sound of] 16 on your cry; when he hears [you he will answer you. Even though the Lord were to give you measured bread and rationed water,] 17 no longer will he hide [from your Master, and your eyes will see your Master.] 18 Your ears will [hear a word at your shoulder which says: This is the path, walk on it,] 19 when you need to go to the right [or to the left. The interpretation of the word, for the last days,] 29 concerns the sin of [...]

Frag. 25 1 [...] the king of Babylon [...] 2 [...] with tambourines and zithers [...] 3 [...downpour and hailstorm, implements of war, they are [...] 4 [...] Blank [...] 5 [Is 31:1 Alas those who go down to] Egypt! In horses [they trust and they rely on chariots] 6 [beca]use they are numerous, and on cavalry, because they are very strong, [without regard for] 7 [the H]oly One of Israel or [consulting YHWH. Blank] 8 [Its interpretation: they] are the people which relies [...]

Frag. 26 1 [Is 32:5-6 No longer] will they call the fool an aristocrat, or treat the rogue as an aristocrat. For] 2 [the rogue says] roguish things. [His heart plots crime; he commits evil and speaks deceitfully against] 3 [the Lord, leaves the hungry person empty and takes water away from the thirsty...]

4QIsaiah Pesher (4Q164 [4QpIsd])

Frag. 1 1 [he will trea]t all Israel like {jet} around the eye. Is 54:11 And your foundations are sapphires. [Its interpretation:] 2 they will found the council of the Community, the priests and the peo[ple...] 3 the assembly of their elect, like a sapphire stone in the midst of stones. Is 54:12 [I will place] 4 all your battlements [of rubies]. Its interpretation concerns the twelve [chiefs of the priests who] 5 illuminate with the judgment of the Urim and the Thummim [...with-out] 6 any from among them missing, like the sun in all its light. Is 54:12 And a[ll your gates of glittering stones.] 7 Its interpretation concerns the chiefs of the tribes of Israel in the las[t d]ays [...] 8 of its lot, their functions [...]

4QIsaiah Pesher (4Q165 [4QpIse])

Frags. 1-2 1 ... [...] ... 2 and Jerusalem [...] And what is written: [Is 40:11 {He carries them on his chest and leads the mothers}.] 3 The interpretation of the word [concerns the Teacher of Righteousness who] reveals just teaching [...Is 40:12 Who has measured the sea in fistfuls,] 4 or [charted] the sky [in palm-breadths, or the dust] of the earth [by bushels. Who] has weighed [the mountains on the balance or the hills in the scales?]

Frag. 5 1 Is 21:9-10 [and all the statues of their gods has he smashed] to the ground. [My people, threshed on the threshing floor, what I have heard from the Lord of Hosts, God] 2 [of Israel, I will tell you.] The interpretation of the word concerns [...] 3 [Oracle against Dumah: Is 21:11-15 Someone sh]outs from Seir: Watchman, what is left of the [night? Watchman, what is left of the night? The Watchman replies: Morning will come and also the night. If you wish to ask, ask,] 4 [come back, return. Oracle against Arabia:] In the scrub of the steppe shall you spend the night, [caravans of Dedan; take out water to meet the thirsty, dwellers in the land of Teman, take bread] 5 [to the refugee, for] / […] he flees in front of the swords / in front of the unsheathed sword / in front of [the taut bow,] / in front of [the fierce fighting. The interpretation of the word concerns…] 6 […] the peoples and the bread […] 7 […] lays waste […]

Frag. 6 1 […] the chosen ones of Israel […] 2 […] eternal. And what is written; [Is 32:5-7 No longer will they call the fool an aristocrat,] 3 [or] treat [the rogue] as superior. For the r[ogue says roguish things and his heart is dedicated to evil, to commit wicked deeds] 4 [and to speak] absurdities against [the Lord]; to destroy [the soul of the hungry person and take water away from the thirsty. As for the rogue] 5 [his roguish deeds are] illicit and he hatches plots [to destroy the poor with lies] 6 [and the helpless who defends] his rights. Its interpretation concerns […] 7 […] … to the law […] … […]

Habakkuk Commentary (Pesher Habakkuk, 1QP HAB)

The scroll, which contains 13 columns of Hebrew writing, consists of two pieces of soft leather sewn together with linen thread between columns 7 and 8. The columns are about 10 centimetres wide; the scroll was originally about 160 centimetres long. The first two columns, however, are badly mutilated, as is also the bottom of the scroll; this produces an undulating break. along the bottom when the scroll is unrolled. The present maximum height of the scroll is 13.7 centimetres; originally it may have been 16 centimetres high or more. Palaeographical estimates of the age of the scroll vary by some decades, but a date around the middle of the first century B.C. or shortly afterwards is probable.

The scroll contains the text of the first two chapters of Habakkuk. The book of Habakkuk, as we know it, consists of two documents: (*a*) 'The oracle of God which Habakkuk the prophet saw' (chapters 1 and 2), and (*b*) 'A prayer of Habakkuk the prophet, according to Shigionoth' (chapter 3). Our scroll quotes one or several clauses from the former document, and supplies a running commentary on the words quoted; but it does not contain the text of the second document, nor, does it make any comment on it. It is plain from the scroll that it never reproduced or expounded the third chapter of Habakkuk, for the original ending is clear for all to see. The omission of all reference to the 'prayer of Habakkuk' is not due to any idea that such a psalm was unsuitable material for a commentary of the kind that is supplied for the 'oracle' of Habakkuk (commentaries of this kind on the Psalter and other biblical poems have been found at Qumran); it is due, more probably, to the fact that Habakkuk's [p.6] 'prayer' was considered to be a separate work, quite distinct from his 'oracle'. After quoting a section of the text of Habakkuk, our commentator says: 'Its interpretation concerns…'—and then proceeds to give its meaning as he sees it, mainly in terms of persons and events of his own time, or of the times immediately preceding and following his own. The Hebrew word rendered 'interpretation' here is *pesher*, and from its frequency and distinctive usage in this commentary, it has come to be used of the commentary as a whole and of others belonging to the same class. Quite a number of such *pěshārîm* have been found in the Qumran caves, but this commentary on Habakkuk is not only the first to be known, but it is the most complete of those that have come to light thus far. It is, besides, of more than ordinary interest because it remains our chief source for some of the most fascinating problems of Qumran study—the character and identity of the Teacher of Righteousness (the founder and leader of the Qumran community), and his relations with various opponents, such as the Wicked Priest, the house of Absalom, the Man of Falsehood and the Seekers after Smooth Things; together with the identity of the Kitti'im, the brutal Gentile power whose domination of Judaea is regarded as a divine nemesis on the wicked rulers of the land. The *pesher* which this commentary (like the others of the same class) provides for the biblical text is an interpretation which cannot be reached by man's unaided wisdom; it is given by divine revelation. A problem which can be solved only with divine aid is evidently no common problem; it is, in fact, a mystery conceived in the mind of God.

Page 1

1. ... shall I cry and not (part of verse 1:2)
2. ... this generation
3. ... oth (fem pl.) upon them
4. ... his about me
5. ... m ... iyt ... (a portion of 2 words in verse 1:3)
6. ... to (or God) by exploitation and from
7. ... (fragmented word) and many
8. ... (2 fragmented words) he
9. ... therefore the Torah fades away (portion of verse 1:4
10 ... which is his burden in the Torah of God
11 ... (direct object) the righteous
12 ... He is the Moreh ha-tsedek
13 ... therefore the judgement goes out (part of 1:4)
14 ... and not the
15 ....

Page 2

1. it be told him ... the treacherous with a man (1:5)
2. all the liars ... the Moreh Tsedek from the mouth
3. of God and concerning the trai(tors) ... the new (fragmented word)
4. our belief in the covenant of God ... those possessing ... and
5. truth. Pesher of the word ... ..gadiym to the last (aleph)
6. days. They are aro... ... ..oth which they will not believe
7 when they hear all the b... ... of the final generation from the mouth (1:5)

8. of the priest which God gives in ... to explain to all
9. the words of His servants coming ... hands God tells
10. all coming to his people and. ... (be)cause behold I am raising up (1:6)
11. the Chaldeans, the nation the bit(ter) ... ... (the has)ty (1:6)
12. Pesher about how the Romans a... ..h speedy ones and mighty men
13. in war to destroy r.. .... from the government of
14. the Romans to do evil ... ... and we do not say
15. according to their laws ...
16. ....

Page 3

1. and in contending and going to strike and to loot the cities of the land
2. because as he says: to possess inhabited areas that are not theirs. (1:7) Dreadful
3. and terrible they are and their judgement and pride proceeds from themselves. (1:7)
4. Pesher about the Romans by which their fear (...) upon all
5. nations and in fact all their thoughts to do evil and (...) to defraud
6. and to come with all the peoples. Their horses are lighter than leopards and more alert (1:8)
7. than the evening wolves. Their horsemen spread themselves from afar 1:8,9)
8. and they fly as an eagle hastening to eat. (1:8) All of them coming for violence multiplying(1:9)
9. by their faces the east (wind). Pesher about the Romans who
10. possess the earth with horses and with their beasts and extend themselves
11. and they come from the isles of the sea to destroy ... and (..)iyl the peoples as an eagle
12. and they are not sated and in them and their beasts ... and even their raging
13. faces and they speak with ... ... as he
14. says: ma(gemet) .... as sand the captivity (1:9)
15. ....

Page 4

1. they shall mock (kings) and laugh at the princes (part of 1:10) Pesher that
2. they mock about many and despise the glorified kings
3. and rulers and they deceive and scorn many peoples. And he
4. (scorns) every fort and laughing at them he heaps up dust and captures them (part of 1:10)
5. Peshru about the rulers of the Romans who despised
6. the forts of the peoples mocking and laughing at them
7. and at many other peoples, imitating and making them foolish to hold them in fear
8. and take them in their hand to destroy them because of the sins of those dwelling
9. there. Then shall he change his spirit and he shall pass over attributing this, his power (most of 1:11)

10. to his god (end of 1:11) Peshru about the rulers of the Romans
11. who by the counsel of the house of Ishm(ael) and the servants of men
12. who rule in the presence of their neighbor ...afterward and they go
13. to slay the .... his strength to his god. (part of 1:11)
14. Peshru about .... (...)l the peoples
15. ....

Page 5

1. you have set them for judgement O Rock, you have established them for correction. (You have) purer eyes(parts of 1:12 and 13)
2. than to view evil and you are not able to consider wickedness (part of 1:13)
3. Pesher about the saying that God is not able to (to save) his people from the hand of the Gentiles
4. and God gives judgement in the hand of his "Chosen one" (to) all the Gentiles and in their reproaches
5. and their guilt to all those doing evil to my people which keep my commandments.
6. in and enemy to whom because that which says you are of purer eyes to behold
7. evil (part of 1:13) Peshru that their eyes do not hunger after the evil end (time).
8. Why do the traitors look and plough evil swallowing (part of 1:13)
9. righteousness away from us. (part of 1:13) Pesher about the House of Absalom
10. and the men of their council which fall silent in reproaching of the Moreh Tsedek.
11. But they do not help him against the lying man who does violence to
12. The Torah among all the nations and makes men as the fish of the sea (part of 1:14)
13. as creeping things to watch over it. (part of 1:14) All of them take up into their net (part of 1:15)
14. and they gather them ..... in their seine. Therefore they are happy (part of 1:15)
15 .... their portion.

Page 6

1. The Romans, the ones gathering wealth from all the spoil
2. as fish of the sea, wherefore he says: (Begin 1:16) Therefore they sacrifice to their net
3. and offer incense to their seine (end part of 1:16) Peshru about them
4. sacrificing to their symbols and all the wars they
5. are showing because in them they fatten their portion. They eat from my table
6. Pesher about their dividing of the world and
7. their violence in devouring the peoples year after year
8. destroying with the sword many lands. (begin 1:17) Therefore they draw out their sword and continue
9. to murder the nations without compassion (end 1:17)
10. Peshru about the Romans who destroy many by the sword

11. youths, mature men, old men, women, infants, and on the fruit
12. of the womb they have no mercy. (begin 2:1 Let me stand fast on my watch tower
13. and present myself on the fortress and I will keep watch to see what He will say
14. to me and w(hat ... I will answe)r when he corrects me (end part of 2:1) (begin 2:2) And YHWH answered me
15. (and said write the vision and make it clea)r on tablets so that the one runn(ing) (end part of 2:2)
16. ....
17. ....

Page 7

1. And God said to Habakkuk to write the things to come
2. upon the last generation but he did not reveal to him the close of the end time.
3. And when he says: so that the one running may read it. (verse 2:2)
4. Peshru about the Moreh Tsedek who makes known to
5. all the secrets in the words of his servants the prophets. Because he will yet have vision(part of 2:3)
6. for a season and shorten the end (time) and he will not lie.(2:3)
7. Pesher about his lengthening the "end time" and that which remains concerning all
8. that which the prophets spoke because God secretly acts to remove the infection.
9. If he delays, wait for it, it will surely come and not be (part of 2:3)
10. late. (part of 2:3) Pesher about the "men of truth"
11. who are practicing the Torah who do not slack their hand from serving
12. the truth but continue to raise up to them the end time because
13. all our hope is that he will bring the establishment according to the Statutes
14. to them as in their secret stock pile. Behold his exalted (soul) is not straight (part of 2:3)
15. (soul in lacuna) ..... Peshru that which also they multiply to them
16. .... their land in judgement ....

Page 8

1. Pesher about all the ones doing the Torah in the house of Judah by which
2. God delivered them from the House of Judgement in serving their evil doings and their Faith
3. in The Moreh Tsedek. (begin 2:5)And so, because he is treacherous through wine, he is a proud warrior and has no
4. place of abode, while enlarging his inner desires as Sheol and is as death and never satisfied
5. while he gathers all the nations and accumulates populations to himself. (end 2:5)
6. (begin 2:6) Shall not these lift up a proverb against him and a mocking riddle for him
7. and they shall say, Woe to him that gains increase from that which is not his,

How long will he forcefully
8. make debtors to himself? (end of 2:6) Peshru about the wicked priest who
9. is called by the name of "The Truth" He stands in prayer while watching
10. over Israel. His heart is lifted up yet he abandons God and uses treachery by their laws for the sake of
11. wealth. He rejoices and he gathers the wealth of the men of violence who rebel against God.
12. And he receives the wealth of nations to add evil to himself by the guilt of his own way
13. ... working in all the corrupt wanderings. (begin 2:7) Shall not those who will exact usury on you arise
14. suddenly and those terrorizing you shall awaken and you shall be for plundering to them. (end of 2:7)
15. Because you have looted many nations the remainder of the peoples shall loot you;(part of 2:8)
16. (Pesher about) the priest who rebels.....
17. ....

Page 9

1. a plague My judgement on evil and lessons from dreams
2. evil (dreams) which they make against me and they are established among the nations by their looting wherefore
3. he says: (begin 2:8) Because you have looted many nations they shall loot you, all
4. that will remain of the peoples (end part of 2:8) Peshru about the priests of Jerusalem
5. The final (end time) ones who gather up wealth and take a cut from the spoils from the peoples
6. and for the "Last Days" they give their wealth with spoil into the hands of
7. The Roman army. (spatium) Because they (the spoils) are that which remains of the peoples
8. (begin the rest of 2:8) from the blood of mankind and violence of land, city and al(l) those dwelling in it. (end 2:8)
9. Peshru about the "Wicked Priest" who (is given) into the hand of the Moreh
10. Tsedek and the men of his council. God gives into His hand his enemies to answer him.
11. (He visits them) with a plague in the bitterness of his soul because they had done wickedly
12. against his "Chosen." (begin 2:9) Woe to the one who is gaining an evil advantage for his household to set
13. his nest on high so he can be delivered from the hand of evil. (end 2:9) (begin 2:10) You have devised shame
14. on your own house, and by cutting off many nations you are sinning against your own (so)ul. (end of 2:10) (begin 2:11) Because
15. a st(one from th)e wall (of your house) will cry out and a beam from the timber

will a(nswer it.) .....
16. ....upon the .....which.....................

Page 10

1. to be exploiting a stripe on the spine (idiom for weakness?) with plunder wherefore
2. he says: (begin part of 2:10) cutting off many nations and sinning against your own soul.
3. Peshru that is the "House of Judgement where God gives
4. his judgement in the midst of many peoples and by name brings us up to judgement
5. And in the midst of them he saves us and by sulphurous fire he judges us. (begin 2:12) Woe
6. to the one who builds a city in blood and founds a town by evil. (end 2:12) (begin 2"13) Is it not
7. Behold of YHWH of Hosts that the peoples should toil in a great deal of fiery trials part of 2:13)
8. and the population tire themselves out with a great deal of trivia. (end 2:13)
9. Pesher of the word about the flying of the liar who leads many astray
10. to the daughters of the city of vanity and by blood and to raise up a city by a lie.
11. For the sake of (self) glory they lead many to serve vanity and for their own profit
12. in b..shey lies to possess ills to empty so that they will be led
13. to the judgement of fire which they blaspheme and shame the "Chosen One" of God
14. (begin 2:14) Because the earth shall be filled with knowing the Glory of YHWH as the waters
15. cover the s(ea.) (end 2:14) ..... Pesher about the word....
16. and their return ....
17. ....

Page 11

1. the liar. And afterward they roll to themselves evil as waters
2. of the seas heaped up. (begin 2:15) Woe to the one causing his neighbor to drink joining your bottle
3. to him and even making him drunk in order to look on their nakedness. (end 2:15)
4. Peshru about the "Wicked Priest" who
5. pursues after the "Moreh Tsedek" to swallow him up in rage
6. his hot (rage). Tenderness is to come to pass and in the last season (end time) a comforting
7. day, The young lions shall appear to them to swallow them
8. and for their stumbling. In the day of fasting of the Sabbath of their comfort. (begin 2:16) You are filled up
9. with shame instead of glory; you are drunken, even you, and are uncircumcised

10. and the cup of YHWH's right hand is turned on you and utter disgrace
11. upon your glory. (end 2:16)
12. Peshru about the "Priest" who is more proud of his shame than of his glory
13. Because he is not circumcised with the circumcision of his heart and he goes on his way
14. for the sake of his profit so that the telling of the fastings will cover them.
15. ....
16. .... to swallow us l(.....) .....

Page 12

1. shall cause terror because of human blood and violence of the land, city and those who dwell in it. (part of 2:17)
2. Pesher of the thing about the "Wicked Priest" to take the spoils to himself
3. as recompense which completes his lust because the moon is his
4. singular witness and their beasts are the fools of Judah doing the
5. the Torah with which God will judge us in the end.
6. wherefore he says: (begin part of 2:17) from blood
7. city and violence of the land, Peshru: The city is Jerusalem
8. in which the "Wicked Priest does abominations and defiles
9. the Holy Temple of God. And he does violence to the earth (part of 2:17) that is to the cities of Judah who
10. lusting plunder wealth. (begin 2:18) What does an idol profit the maker of it who fashioned it
11. a molten image that the one fashioning the work trusts in it
12. to make mute gods. (end 2:18) Pesher concerning the word about all
13. of the idols of the Gentiles which they have made for themselves to serve
14. to them, and they will not deliver them in the day of Judgement. (begin 2:19) Woe
15. (to the one saying t)o a tree wake up ...... to a silent (st)one
16..................................

Page 13

1. Be silent before him all the earth (2:20)
2. Peshru about all the nations
3. which serve stone and wood ...
4. in the judgement God will banish all the servants of idols
5. and the evil ones from the earth.

Hosea Commentary (Pesher Hoshe`a, 4Q166 4QpHosa)

This text is a commentary, or "pesher," on the prophetic biblical verses from the book of Hosea (2:8-14). The verse presented here refers to the relation of God, the husband, to Israel, the unfaithful wife. In the commentary, the unfaithful ones have been led astray by "the man of the lie." The document states that the affliction befalling those led astray is famine. Although this famine could be a metaphor, it may well be a reference to an actual drought cited in historical sources of that time.

The manuscript shown here is the larger of two unrelated fragments of the Hosea Commentary found in Cave 4. The script, which is identical to that of a commentary on Psalms, belongs to the rustic, semiformal type of the Herodian era.

The Hosea Commentary Scroll was first published by J. Allegro as the fifth volume of the official publication series, "Discoveries in the Judaean Desert." In 1979, M. Horgan completed a work on all the "pesharim," or commentaries, which included an extensive treatment of the Hosea Commentary fragments. The "pesharim" interpreted the biblical text in light of events of the late Second Temple Period--seeing within the text prophesies and messages relevant to the community's beliefs and practices. Copied late first century B.C.E.
Height 17.5 cm (6 7/8 in.), length 16.8 cm (6 5/8 in.)

4Q166 (4QpHosa)

Hos. 2:10-14

1. (10)[SHE DID NOT KNOW THAT] I MYSELF HAD GIVEN HER THE GRAIN [AND THE WINE]
2. [AND THE OIL, AND] (THAT) I HAD SUPPLIED [SILVER] AND GOLD ... (WHICH) THEY MADE [INTO BAAL. The interpretation of it is]
3. that [they] ate [and] were satisfied, and they forgot God who [had fed them, and all]
4. his commandments they cast behind them, which he had sent to them [by]
5. his servants the prophets. But to those who led them astray they listened, and they honored them [ ]
6. and as if they were gods, they fear them in their blindness.
7. vacat
8. (11)THEREFORE, I SHALL TAKE BACK MY GRAIN AGAIN IN ITS TIME AND MY WINE [IN ITS SEASON,]
9. AND I SHALL WITHDRAW MY WOOL AND MY FLAX FROM COVERING [HER NAKEDNESS.]
10. (12)I SHALL NOW UNCOVER HER PRIVATE PARTS IN THE SIGHT OF [HER] LO[VERS AND]
11. NO [ONE] WILL WITHDRAW HER FROM MY HAND.

12. The interpretation of it is that he smote them with famine and with nakedness so that they became a disgra[ce]
13. and a reproach in the sight of the nations on whom they had leaned for support, but they
14. will not save them from their afflictions. (13)AND I SHALL PUT AN END TO ALL HER JOY,
15. [HER] PIL[GRIMAGE,] HER [NEW] MOON, AND HER SABBATH, AND ALL HER FEASTS. The interpretation of it is that
16. they make [the fe]asts go according to the appointed times of the nation. And [all]
17. [joy] has been turned for them into mourning. (14)AND I SHALL MAKE DESOLATE [HER VINE]
18. [AND HER FIG TREE,] OF WHICH SHE SAID, "THEY ARE THE HIRE [THAT MY LOVERS HAVE GIVEN] ME."
19. AND I SHALL MAKE THEM A FOREST, AND THE W[ILD BEAST OF THE FIELD] WILL DEVOUR THEM.

Zephaniah Commentary (4QZephaniah Pesher 4Q170 a 1:12-13)

1 The Lord will not do good or evil. Their wealth will be plundered and their houses will be destroyed. Its meaning is . . .
2 . . . they will not eat . . .
3 What is says: their wealth will be destroyed . . .
4 Its meaning is . . .

Micah Commentary (4QMicah Pesher, 4Q168)

(a commentary on Micah 4:8-12)

1. . . . of the daughter of Jerusalem. Now, why do you cry out? Do you not have a king among you? Have you lost your advisor? 2. Have labour pains gripped you like a woman giving birth? Have labour pains, daughter of Zion, like a woman giving birth, for 3. now you will come out of the city and settle in open fields, and you will go to Babylon. There you will be saved. There 4. the Lord will deliver you from the hand of your enemies. Now, many nations will assemble against you, 5. saying, let her be defiled and let us set our eyes upon Zion. They do not know the thoughts 6. of the Lord, and they do not understand his plan . . .

The Nahum Commentary (4QNahum Pesher 4Q169)

Prophetic commentary which alludes to the history of the sect and names real historic figures, such as Alexander Jannaeus, "the Lion of Wrath." Composed in the 2nd Century BCE, This copy from the 1st Century CE.

1 Whither the lion, the lioness, the lion's cub
2 [and none to terrify, Its interpretation concerns Deme]trius King of Greece, who sought to enter Jerusalem by the counsel of the Seekers after Smooth Things
3 [and God did not surrender Jersualem into] the hands of the Kings of Greece from Antiochus until the rise to power of rulers of the Kittim; but afterwards (the city) shall be trodden down.
4 [ ] The lion tears sufficient for his cubs and strangles prey for his lioness
5 [Its interpretation] concerns the Lion of Wrath who strikes by means of his great men, and the men of his counsel
6 [he filled his cave] his cave and his den with torn flesh. Its interpretation concerns the Lion of Wrath
7 [Who   sentence of] death on the Seekers after Smooth Things and who hangs men alive

Targum of Job (4QtgJob, 11QtgJob)

Two targum manuscripts of Job were found at Qumran. Since they are both incomplete and do not overlap, we do not know whether they represent the same targumic tradition or not, although that is likely. The Aramaic Text is relatively straight forward, with an occasional tendency to abridge slightly. Extant portions contain none of the flights of midrashic expansion that one gets in some other targumic traditions.

Text

4Q Targum of Job (4Q157[4QtgJob])

Frag 1 col.i
(= *Job 3:5-?*) $_2${...} a cloud[will come] over him $_3$ [... in a time not in the same dimension $_{4-5}$ [...]...
Frag 1 col.ii
(=*Job 4:16-5:4*) $_1$...[...] $_2$ Can a man speaking to God [be rigth?...] $_3$ and to his angels[he causes insanity...] $_4$ which [are formed] in dirt[...] $_5$ and many [...] die and not from knowledge[...] will you contemplate? *Blank* Maybe he does not kill the weak[...] $_8$ But I have seen a cruel person ...[...] $_9$ ...[...].

Fernando Klein

11Q Targum of Job(11Q10[11QtgJob])

Col. I
(=*Job 17:14-18:4*) ₁ [... my mother and sister are dead]. And what did I do [...]₂[...] Maybe [they will go] with me to Sheol?[...] ₃ [... in the dirt] we will be buried? Blank ₄ Bildad the Shu[ite replied...] ₅ [...] will you complete you thought? [...] ₆ [...] do we look like animals?[...] ₇[...] Maybe from your view point [...] ₈ [... the rock] from its point?[...]

Col. II
(=*Job 19:11-19*) ₁ I bore his rage and [thinks I'm...] ₂ His henchmen came and destroy [... My brothers and I] ₃ have remourse, my house staff. My butler, does not recognize [...] I call but he does not respond [...] ₆ I'm embarrassed to show myself to my wife [...] ₇ The evil hurts me [...] ₈ everyone who[...]

Col. III
(=*Job 19:29-20:6*) ₁ [...] evil. Blank [...] ₃ [... he answered: Here is my heart [...] ₄ [...] I will listen to my crimes, but the soul [...] ₅ [... Do you not understand infinity, from ...[...] ₆ [...] Because praising the cruel [...] ₇ [...] goes by quickly [...] ₈ [...] and he looked [toward] the sky[...].

Col. IV
(=*Job 21:2-10*) ₁ [...] personally[...] I know you laugh. [...] ₃ as a result [my soul]does not get [tense...] ₄ be quiet...] ₅ I am mesmerized. Why do the corrupt become richer? Their children[...] ₇ in plain sight. Their houses [...] ₈ God is with them. [...] ₉ their (cow) gives birth[and does not miscarriage..].

Col. V
(=*Job 21:20-27*) ₁[...] he looks [...] their destruction and around[...] ₂ [...] would like God in their home[...] [...] ₃[...] life is short ? is God [...] ₄[...] the most powerful? His assistants [...] ₅ the care of the bones. Another dies [cruelly] in spirit [...] ₆ [...] starving, they both[lie on the dirt...] ₇ [...] on top of them I know [what your thinking...] ₈ [...] you planned [against me][...]

Col. VI
(=*Job 22:3-9*) ₁ [...] God ₂ [...] your way [...] will he make a convenient with you? ₄[...] there is no ₅ [...] your brothers for nothing ₆ [...] the parched not ₇ [...] bread. And you said ₈ [...] his face ₉ [...] was emotionless.

Col. vii
(=*Job 22:16-22*) ₁ they passed away[...] ₂ They prayed to G[od..] ₃ to our God [...] But the evil group [...] ₅ and marked and [...] ₆ How can that [...] not [...] ₇ Look [...] ₈ Receive [...].

Col. VIIa
(=*Job 23: 1-8*) ₁ [...] Job replied [...] ₂[...] because my voice [...] ₃ [...] my whining. Indeed, I would know and I would find God ₄ [...] heaven. I would say to[God ...] ₅ [...] I would repent an I would know [...] ₆ [...] I know what he will say to me. [...] ₇ [...he might] treat me unfairly. Indeed until [...] ₈[...] for truth and how [...] ₉ [...] If advancement[...].

Col. VIII
(=*Job 24: 12-17*) $_1$ From cities[...] $_2$ he complains $<_4$ in its trail [...] $_5$ and to the needy; and in the evening. [...] $_6$ the darkness saying[...] $_7$ and he will sin [...] $_8$ in evil[...] $_9$ for them[...]

Col. IX
(=*job 24:24-26:2*) $_1$ [...] they come together $_2$ [...] Who will answer me and [...] $_3$ [...] Blank Bildad replied [...] $_4$ [...] God controls everything; he does [...] $_5$ [...] in his power. Is there trust for[...] $_6$ pr for whom does [...] not rise $_7$ [...] God and how will he be fair [...] $_8$ [...] unchanged and the galaxy [...] not $_9$ [...] mankind, this warm [...] $_{10}$ [...] and he said << Can you possibly,[...]?

Col. X
(=*Job 26:10-27:4*) $_1$ [...] to the realm of evil; $_2$ [...] he dissects them and they are worried about $_3$ [...] the sea, and he killed with his knowledge. $_4$ [...] he makes I glimmer, his hand struck fleeing snake. $_5$ [...] their paths. And we only hear and echo. $_6$[...] he will know>> Blank $_7$ [...] Blank $_8$ [...] and said $<_9$ [...] my spirit which while [...] $_{10}$ [...] in my nose they won't say [...]

Col. XI
(=*Job 27:11-20*) $_1$[...] in God's control and the work of $_2$[...] can be viewed by all. Why $_3$ [...] the cruel man $_4$ [...] they steal from him. If $_5$ [...] the sword , they will plunder, and feel fulfilled $_6$ [...] and their [...] no $_7$ [...] money, and increases like dust $_8$[...] and honest man will give away his wealth $_9$ [...] like a house $_{10}$ [...] lies down and is not taken. $_{11}$[...] like water the evils.

Col. XII
(= *Job 28:4-13*) $_1$ foot[...] $_3$ sapphires [...] $_4$ not [...] $_5$ the serpent enters. [...] $_9$ man [...].

Col. XIII
(=*Job 28: 20-28*) $_1$ the place of knowledge? [...] $_2$ it hides from the birds of the sky. [...] $_3$ << By word of mouth we know who you are >> [...] $_4$ in it, since he [...] $_5$ for reaching [...] $_6$ When he made the wind [...] $_7$ by one fall stroke. When he made [...] $_8$ fluffy clouds. Meanwhile[...] $_9$ And he said to the sons [of mankind...] $_{10}$ and to leave from [...].

Col. XIV
(=*Job 29: 7-16*) $_1$ in the morning at the entrance of the city in the center of town[...] $_2$ Children, when they see me and [scholars] $_3$ Powerful men don't speak to me and push me aside [...] $_4$ The leaders disguise their voices;[...] don't speak. $_5$ They once praised me when I spoke[...] because I freed the poor [...] $_7$ no one helps me. The blessing of the last one [...] $_8$ the widow prays for me [...] $_9$ I wore a garment made of goats skin [...] $_{10}$[...] and feet for the lame [...] $_{11}$[...] I did not know[...].

Col. XV
(=*Job 29: 24-30:4*) $_1$ [...] I thanked them and they did not believe[...] $_2$ [...] I chose my way and I was in control [...] $_3$ [...] at the top of his legions, an like a man who [...] the depressed $_4$ [...] They harassed my children [...] $_5$ [...] Their fathers would not sit with the lower class.[...] $_6$[...] I did not like them and under their influence [...] $_7$ [...] they searched for food to feed their soul[...] $_8$ [...] evil which they ate [...] $_9$[...] sticks as their bread [...]

Col. XVI
(=*Job 30: 13-20*) $_1$ [...] they came to destroy me, and there is no savior. $_2$ [...] for them. As I become even greater in pain $_3$ [...] The evil pain trys to over come my body $_4$ [...] I have no possessions $_5$ [...] my salvation. Now the pain irritates me $_6$ [...] days of intense pain I feel $_7$ [...] my bones and joints ache tremendously[...] $_8$ [...] I thrash around in pain $_9$[...] they encircle me and throw me to the ground $_{10}$ [...] to you [...]

Col. XVII
(=*Job 30 :25-31:1*) $_1$ [...they] harassed [me] and not $_4$ [...] I walked $_5$[...] I shouted $_6$ [...] for the ostriches $_7$ [...] of

Col. XVIII
(=*Job 31 :8-16*) $_1$ He will eat [...] my heart for a woman [...] $_3$ She will smash [...] anger $_4$ and is a sin[...] which up to Abaddon shall eat[...] If I was quick in judging my servant [...] what will I do $_7$ when he awakes [...] look $_8$ he made me [...] oneself. If I denied [...] I stopped to be consumed.

Col. XIX
(=*Job 31 :26-32*) $_1$ It was visible, and at the moon [...] my heart, $_2$ and kissed my mouth[...] I would have lied $_3$ to El Shadi[...] I become happy $_4$ in his misfortune [...] my plaque, and he listened [...] in my rage $_6$ and took [...] my taste of sin by asking [...] the men $_8$ of my house: who [...] $_9$ [did] not [...]

Col. XX
(=*Job 31:40-32:3*) $_1$ substituted for wheat [...] $_2$ [...] from the pine. Completed are [...] $_3$ These [...] from answering [...] Job was honest[...] $_5$ Blank $_6$ Meanwhile he became angry [...] of the clan of Rome[...] $_8$ and also against [...] $_9$ words [...]

Col. XXI
(=*Job 32: 10-17*) $_1$ my words, I as well. O.K. I waited [...] $_2$ you stopped, though you wanted to finish[...] $_3$ and you gave Job nothing [...] $_4$ to his knowledge. Maybe you should say [...] $_5$ for this we punish God and not man[...] $_6$ words and he does not respond [...] $_7$ and they are quiet while I wait for a response[...] $_8$ they leave and say nothing [...] $_9$ I to said nothing

Col. XXII
(=*Job 33:6-16*) $_1$[...] Alright, my horror will not shock you[...] $_2$ [...] burden. Surely you spoke in my ear and the sound [...] $_3$ [...] I am clean and there is no sin in me, I am blameless[...] $_4$[...] If he finds I have sinned he will take me [...] $_5$ [...] he places me in the prison and binds me with chains[...] $_6$[...] because God is greater than man[...] $_7$[...] you will speak arrogantly, because in all your actions[...] $_8$ [...] God knows how to communicate to everyone [...] $_9$[...] in dreams, during the night [...] while you sleep in bed[...] ...[...] ...[...]

Col. XXIII
(=*Job 33:24-32*) $_1$ and he said<< Free from harm [...] $_2$ from the fire that consumes him [...] with $_3$ youth and returns to his childhood [...] and he will hear him $_4$ and will see his face when saving him[?...} and based on his work he will reward him. And he will say [...] yet $_6$ he has not rewarded me based on my choice. He has saved [...] $_7$ It will be clear. Behold [...] $_8$ [on]ce, twice, three times[to the] man for [...] $_9$ living (creatures). Be mindful of this [...] I will speak. $_{10}$ [If] you have words [...]

The Dead Sea Jesus

Col. XXIV
(=*Job 34:6-17*) $_1$ of sin. Who [...] sin? And associates $_2$ with criminals [...] cruel men. For he states << A man will change [...] following god>> $_4$ Now, men of [...] God does not deceive or create evil [...] he rewards man $_6$ [...] Will God, possibly, lien now, and lord [...] whom created the earth $_8$ and formed the world? [...] takes air away from him and he will die [...] they shall die $_{10}$ [...] my word. Deception possibly.

Col. XXV
(=*Job 34:24-34*) $_1$[...] to the infinitely powerful , and put other [...] $_2$ [...] he knows their action and throws into the place [...] $_3$[...] his way and have not kept this covenant[...] $_4$[...] of the poor and hears the cry's of the tortured [...] $_5$ [...] covers his face who will answer him about a tribe [...] $_6$ [...] the evil man has control. They create [...] trip. $_7$[...] I pray for him, in only him [...] $_8$ [...] I did not pursue, because [...] $_9$ [...] you decide and not I [...] $_{10}$ [...] words and man[...]

Col. XXVI
(=*Job 35 :6-14*) $_1$ to you. And at a time your wrong doings rise, how do you [handle him?] Are you right, what $_2$ does he need, or what does he get in return? Your wrongs (change) [a human similar to yourself] $_3$ your equally, a child of mankind. As a result of the multitude [of enslavers] they cry and wail $_4$ facing all; yet they don't [ask where is] God $_5$ who created us and gave us [...] for farming $_6$ at night; who separated us from animals and has made us smarter than birds? $_7$ They cry, yet he does not [respond out of righteousness] $_8$ to the sinner. For God [does not hear those who mislead, and the lord to] the insignificant, shows no interest. If you say [...] $_{10}$[...] ...[...]

Col. XXVIII
(=*Job 36:23-33*) $_1$ you [achieve inequality. Knowing] that their actions are right, men have witness these actions. Every man considers them and the children of man kind view from a distance. God is all powerful and immortal.[wee do not] know [them] , and how long he lives. For $_5$ [he tracks the] clouds and directs the rain storms, and their clouds release $_6$ [rain drops] [upon] many people. Indeed who molds the clouds[with great din] who hides and reveals[light] [...] hidden ; he will use them to judge the nation, $_9$ [...] upon his command [...] $_{10}$[...] control them[...]

Col. XXIX
(=*Job 37 :10-19*) $_1$ above the water. Using water he causes the cloud to discharge fine And he says<< The people should here this!>> and they perform their jobs; he placed the people in control of everything on land. Whether to help or destroy or for starvation and poverty . Consider this Job, and rise contemplate the power of God. Have you any idea what God has placed upon them, and how he makes light shine from clouds? Can you protect the cloud with your powers? Since your power [...] Because he has infinite knowledge[May be you create] the storm clouds. Can you change a cloud into a mirror $_{10}$ He knows...

Col. XXX
(=*Job 38:3-13*) $_1$ Protect your grain like a man [and I will test] [you] and you will respond $_2$ Where were you when I created the earth? Answer, if you can $_3$ who created , measurements? Or who used a tape measure? Or what are its bases set to or who set the cornerstone. $_7$ When the stars shown in the morning and all the angels

141

of God song? Can you lock the entrace to the sea when it tries to leave the deep murky bottom. When did you where clouds as cloths and fog as baby's cloths. Can you set the limits of the sea. Did you say it can only go this far and not go beyond your waves. In the past did you control [the morning] the ends of the earth [...]

Col. XXXI

(=Job 38: 23-43) $_1$ which[I keep for] times of danger for the day of war and rebellion? [...] where does the wind come from? Does the wind come from the heavens? Who has set the period for rain and a track for the clouds to bring rain to the dessert, where no man lives; to water the plains to cause grass to grow. Who is the father of the rain, and who controls the fog. And who produces the frost .. . and [darkness of the sky] who created it ?] Like a rock coated with water and the faces of [darkness?] of the Pleiodes or you [open] the fence of Orion[...] you undo the North Star(?) with his sons? [...] $_{10}$ [...] the clouds[...]

Col. XXXII

(= Job 39: 1-11) the goats or birth pangs of [...] they are mature; do you know when they were born. They give birth and the sons become out casts. Do you cause them to leave? They raise their son and force them away. Who set the donkey free and unchained the restraints on the anager? I created the desert as the anager's home and the ground his home and pays no attention to the noise of the city and to the commands of his master. He eats from the mountains grass and eats all that is green. Will the bull choose to serve you or will he sleep in your stable. Will you harness[the bull] with a yoke and will he till the soil behind you. [...] ? Do you trust his strength?

Col. XXXIII

(=Job 39 :20-29) $_1$ [...] Do you scare his (horse) with a powerful [...] $_2$ in his growling fright and fear. He wanders throughout the valley, and shakes and rejoices $_3$ and throws himself into danger. He ignores fear and does not flinch $_4$ from a sword. He prepares to shout and arrow $_5$ as he is armed with a staff and a sword, the bugle sounds and he yells Aha, and from $_6$ a distance he smells combat, and relishes the sound of swords rattling and war cries $_7$ Does the raptor fly with it's wings to the wind? Or does the eagle glide at your command and the $_9$ raptor builds [his] nest high in the cliffs he lives and rests[...] $_{10}$[...] ...[...]

Col. XXXIV

(=Job 40: 5-11) $_1$ [...] end Blank [...] $_2$ God answered Job/ form [out of nowhere(?)/] and the cloud and told him protect your genitals $_3$ then like a man and I will question you and you will answer me Would you assume $_4$ that judgement is void and place blame upon me so you appear innocent? Or $_5$ do you have an arm like God or thunder with a voice like his? Dispose of greatness and haughtiness and wear splendor, in glory and in honor. Dispose of your rage and view the righteous men and humble him and destroy ever $_8$ proud soul and dispose of the rest of the cruel people and bury$_9$ them in the ground Blank and cover them with ashes $_{10}$[...] there is

Col. XXXV

(=Job 40:23-31) $_1$[...] even though] $_2$ the Jordan's banks [should overflow] he trusts that he will receive it [...] $_3$ who will control him when he raises his head, or restrain his jaws. Will you catch a crocodile with a hook or tie a rope around it tongue? Will you put a muzzle on his nose and stab his jaw with a knife. Will he speak ₆ nicely to you or will he speak to humbly? Will he ₇ make a promise with you or will you treat him as a slave for eternity? Will you play ₈ with him like a bird, or chain him up for your daughters? and [...] ₉ ov[er him...] and they shall take him out of [Canaan] ₁₀ [...] of fish[...]

Col. XXXVI

(=Job 41: 7-17) ₁[...] ...[...] ₂ [One] adheres to the other and wind does not flow between them. They ₃ hold each other and they do not separate. His sneeze triggers ₄ the fire between his eyes like the brightness of dawn; from his mouth ₅ torches appear, they leap like tongues of fire; smoke billows from his nostrils, like a torch burning incense; his breath spews coals and sparks leap from his mouth. His neck contains strength and before him ₈ power surges. The fold of his flesh are taunt , forged within him like iron; and his heart [...] like stone [...] ₁₀[...]...[...]

Col. XXXVII

(=Job 41: 25-42:6) ₁[...]...[...] ₂ and he is the king of all reptiles. Blank ₃ Job answered and said to God: I know that you ₄ can do anything, and you do not lack power or wisdom. ₅ I spoke once and I will not revoke it, twice, and ₆ I will not add to it. Listen then and I will say to you; I will question you ₇ and you will answer me. I knew of you only by word of mouth and now I have seen you for this I will be obliterated and destroyed and will turn into dust ₉ and ash Blank

Col. XXXVIII

(=Job 42:9-12) ₁ [...] and he did [...] God; and God heard Job's Voice and forgave ₃ his sins on his account. And God turned /to Job/ in his mercy ₄ and doubled all his possessions for him. And there came to ₅ Job all his friends and all his brothers and all his acquaintances and ate ₆ bread with him his house , and comforted him for all the evil that ₇ God had brought upon him. And each one gave him a eve ₈ and each one a gold ring ₉ And God blessed Job in the end, because he had[...]

Fernando Klein

Wisdom Literature

Wisdom Text (1Q26, 4QWisda)

A composition of wisdom and instruction, which Wise, Abegg, and Cook title "The Secret of the Way Things Are," reveals the inflexible purposes of God. Full of words of instruction and consequences, the text reveals how life should be lived. This can be seen in numerous fragments but none are more apparent then 4Q419 frag.1 when paraphrased states "carry out my deeds, following my guidelines". In addition fragment 4Q418 col. 2 reveals, one's actions will be weighed in the palm of His hand. Thus, indicating individuals who live their life following Gods commandments will be allowed into heaven. The consequences for sinners is written clearly in 4Q416 frag. 1, stating "For those who have engaged in sin will be fearful as judgement day approaches and even the darkest depths of Hell will be terrified". Although specific examples of right and wrong are not given, The Secret of the Was things Are gives a general overview of finding personal wisdom and using that to become a righteous human who will make the rectified choices in life.

4Q410 Frag. 1

....$_1$....$_2$[...if you] breach any of the ...$_3$....$_4$ curses will adhere to you $_5$ ...and tranquility will elude you consequently ...$_6$...what is absolutely promising and what is detrimental...$_7$...forever.
   Presently, I, with [the assistance of the Lord] essence...$_8$...he will be truthful ...$_9$ The prophecy regards [...], the dream entails details of the dwelling, for I have envisioned [...]

4Q412 Frag.1

...$_5$ Inflict correction on your tongue, and on your speech. [...]$_6$ Reflect on holy thoughts. [...] to individuals striving [...] $_7$ Constantly glorify [God...] you will quiver [and shiver] $_8$ Exalt His persona...$_9$ the entire congregation...from dusk to dawn [...]

4Q413 Frag.1

Instruction and [intuition] and I will educate you. Now understand the ways of man and the deeds of humanity. He will develop wisdom from His truth, and the refusal of sin, $_3$ [and] shall not be influenced by circumstances seen or heard. And now, $_4$ mercy ...of the founding fathers, and subsequent generations, as God has planned.

4Q415 Frag.6

The elements of man $_2$ You are impoverished, and ...$_3$ you are inadequate in your acquaintance Test these things by the secret of the way things are [...] $_5$ from the source and by the influence.

4Q415 Frag. 9

[Do not let your thoughts reside] $_6$ on Foolishness, do not follow the horde of [evil]...obtain knowledge, for $_7$ by it He created it, for knowledge is the allowance of the [macrocosm]...she constituted them, $_8$ simultaneously, the dominance of man with [women] her soul, dominance is a component of her, for [...] $_{10}$ if one has unequal [...] $_{11}$ In harmony [...men and] women, and in the scheme of [life].

4Q416 Frag.1

[...] and to gauge His desire...$_3$ day by day...$_4$ corresponding to their multitude, [necessary] $_5$ and its empire hear...$_6$ in accordance to the will of their host and the master of the stars He has promulgated $_8$ through their character and attributes $_9$ Between and all their great amount [...] He has counted...
 He will critique evil's deeds, but all those whose hearts resides in truth He will honor $_{11}$ For those who have engaged in sin will be fearful as judgement day approaches.$_{12}$ Even the darkest depths will be terrified, and all the spirits of flesh will strip naked, and the heavenly followers [...] $_{13}$ At the time of justice all malignant deeds will vanish, and the epoch of righteousness will be absolute.$_{14}$ And for infinity, for the He is the God of truth, and for all times, $_{15}$ So one's soul will differentiate between right and wrong [...] $_{16}$ it is the desire of the id, and those who comprehend...

4Q417 Frag.1 Col.1

[Verbalize softly to a leader] $_1$ always, for fear he commands you; one's tone should conform to his manner, [he...] $_2$ without consequences. When congenial, seek, but when argued remain detached. $_3$ do not cause grief, because you utter...quickly relate his criticism, but do not overlook your sins...$_5$ for he is equivalent of you, for he is [...his wishes] he will act, for he is unparalleled in every action... $_7$ of his actions. On the day of judgement, his demeanor will be in accordance, with him [walk].

4Q418 Frag. 77

[...] the secret of the way things are, and master the essence of life and consider at his abilities $_3$ experience has molded him. Only then, can one conceive the pureness of a man's spirit and burden of his [self-discipline] $_4$ his soul, and master the answers of the unknown, the force of the period and the standard of [things to be judged].

4Q418 Frag. 123 Col. 2

[...] $_2$ From the beginning to the end of time...$_3$ all the events that occurred, why things were, the way they were and all lead to an unknown, [future]. $_4$ The time that God exposes, the answers to the unknown, too, those who listen...$_5$ You will be one of the few, to be able to conceive, when I present you with the [answers] $_6$ Your actions will be weighted in the palm of my hand along with time...$_7$ Cherish greatly, the lessons learned [...].

4Q419 Frag. 1
[...] carry out my deeds, following my guidelines [...] $_2$ given to you through Moses, and should be obeyed... Through his ministers, for they poses the [promise] $_4$ Moses will make public which is His and [what] $_5$ From Aaron's descendant's, he will select...$_6$ His paths and to verge on [the flames] which mollifies...$_7$ and He passed on... to those who followed Him $_8$ and He ordered [...] $_9$ the seat of honor exhilarated in splendor [...] $_{10}$ His magnificence, will stand of all time, as will his life...$_{11}$ you will seek, and the abomination of impurity...$_{13}$ In turn for loving, they will wallow...ways.

4Q420 Frag. 1 Col. 2
[...] all must be considered before a response is rendered $_2$ and an answer will not be given, until all is pondered, after fortitude he will respond [humbly] $_3$ he will convey...and will strive to find truthfulness and fairness, and morality $_4$ He will find its roots [...] and his reason will be modest and passive, He will not look back...$_5$ He will not be achieved through manual labor; this will produce a neutralization [within] through sagacity ...] poisons. Stretching to all corners of the world searching for honorable actions.

4Q421 Frag.1 Col 2
Only the knowledgeable and intuitive man, "modest in his stature will chide his teacher, to follow in the foot steps of their God, $_3$ I am the virtuous one..."

Collection of Proverbs (4QWisd)

A Collection of Proverbs also known as the Sapiential Work belongs to the 'Wisdom Literature' of the Dead Sea Scrolls. This cave 4 document is also known as 4Q Wisdom. This simple poetic collection, like any other 'Wisdom' texts, has the usual vocabulary of 'Judgement', 'Riches' and 'Knowledge'. However, there are two viewpoints regarding the format of this collection.

Frag. 1
$_2$ [...] a man [...] $_3$[...] who decides to build a house and covers its walls with plaster. With him too [...] $_4$ the walls of the house will fall down when rain falls on it.
It is not advisable to have any kind of legal contract with a person who is not stable. $_5$ Otherwise, just as a metal like lead that looks intact melts immediately when heated, the unstable person too will change his mind and not keep his word.
$_6$ Do not lay trust on a lazy man to run an important errand for you, because a lazy person will not feel responsible to do the job given to him, do not ask him to fetch something for you, $_7$ because he will not follow the specific orders given to him.
Do not ask a dissatisfied person [...] $_8$ to get any money that you need. It is not

wise to trust a man with a deceitful speech [...] ₉ for he will definitely manipulate your sayings and give a different meaning to your saying and decisions, for he would not care to keep the truth intact. [...] ₁₀ the words that come out of his mouth.

Do not let a stingy man handle money; [...] ₁₁ for he will not remain loyal and may not give back everything that actually belongs to you [...]. ₁₂ and at the time when you need him to repay you, he will turn his face away from you [...] ₁₃ and the short tempered man will for certain cause harm to them. A man [...]

Frag. 2
[...]
Frag. 3
[...]₁ an irresponsible person will not do his work carefully and according to his position or even according to his age. A person who gives his verdict before thoroughly examining the situation, and, a person who believes before looking at the evidence ₂ Do not give him the power to rule over those who seek for Knowledge, ₃ because he will not be able to do justice to his authoritarian position and hence, not being able to understand the judgments of the other wise people under him, he would not be able to distinguish a good man form a wicked person. ₃ So he will also be contempt.

Do not send a man with a vision impairment to observe the upright for [he will not be able to look deep into the situation]

₄ Do not send a man who has a hearing impairment to give his opinion about a dispute and try to solve it, because he would not be capable of solving the problem, like someone who winnows in the wind a grain ₅ that is not completely separated out. It is not helpful when it comes to talking to a ear that is not ready to listen to you or in other words, a biased person, or, a person who lacks the spirit [...]

₆ It is futile to ask a person who is narrow minded or close minded, to give his judgement for he is not willing to accommodate suggestions and opinions from others and hence, his wisdom remains restricted and is not allowed to evolve, ₇ and so he is not able to use his wisdom efficiently. The wise man will be understanding, and he will have the ability to identify wisdom [...] ₈ A man of strong [...] such a person would be zealous [...] ₉ He would argue and fight against those who would deviate from the set rules and principles [...] for the right of the poor of [...]

₁₀ [...] will care for those people who do not have wealth, the children of the good and wise people [...] ₁₁ [...] with all the money of [...]

Wiles of the Wicked Woman (4Q184)

She speaks emptiness and in [...]
She is always looking for mistakes, sharpening the words that come from her mouth, and she flatters men with nonsense and leads them to uselessness.
Her heart sets traps, and her kidneys cast nets.
Her eyes have been invaded by evil, her hands have a tight grip on the Pit.
Her feet come down to do evil and only walk towards crime.
Her thighs are the foundations of the dark, and many sins are under her skirt.
Her [...] are the gloom of night.

Her clothes are dreary night, and her jewelry is drenched in evil.
Her couches are beds of corruption, and her [...] are the ditches of Hell.
Her houses are a home to darkness, she resides within the heart of the night.
She pitches her tents on a foundation of darkness, she rests in the tents of silence, amidst the everlasting flames.
She does not associate with those who shine.
She is the beginning of all paths to evil.
She will ruin all those who possess her, and destruction will come to all those who take hold of her.
Her paths are the paths of death, and her ways are roads to sin, her trails lead toward wickedness, and her pathways, to evil wrongdoing.
Her doors are the doors of Death, and in through her doorway is Hell.
Those who enter there will never return, and those who partake of her will fall into the Pit.
She hides in secret all [...].
She disguises herself in the city streets, and she plants herself by the city gates.
No one will keep her from her never-ending fornication.
Her eyes dart here and there, looking for a virtuous man to catch, an important man to lead astray, a just man to make unjust, to draw the righteous from obeying the commandments, to bring the good man down, to cause the honest to break the law.
She causes the meek to rebel against God, and turn their steps away from justice, to put vanity in their hearts so that they do not stay on the path of righteousness.
She seeks to lead men to the paths of the Pit, to flatter the sons of men with smooth words.

The Parable of the Bountiful Tree (4Q302a)

Frag.1 Col.2

Please consider this, you who are wise: If a man has a fine tree, which grows high, all the way to heaven (...) (...) of the soil, and it produces succulent fruit every year with the autumn rains and the spring rains, (...) and in thirst, will he not (...) and guard it (...) to multiply the boughs (?) of (...) from its shoot, to increase (...) and its mass of branches (...)

Frag.2 Col.1

(...) your God (...) your hearts (...) (...) with a willing spirit. (...) Shall God establish (...) from your hand? When you rebel, (...) your intentions, will He not confront you, reprove you and reply to your complaint? (...) As for God, His dwelling is in heaven, and his kingdom embraces the lands; in the seas (...) in them, and (...)

Fernando Klein

Messianic and Visionary Works

The Messiah of Heaven and Earth (4Q521)

This text is one of the most beautiful and significant in the Qumran corpus. In it many interesting themes that appear in other Qumran texts reappear. In the first place, there is continued emphasis on 'the Righteous' (Zaddikim), 'the Pious' (Hassidim), 'the Meek' ('Anavim), and 'the Faithful' (Emunim). These terms recur throughout this corpus (in particular see the Hymns of the Poor below) and should be noted as more or less interchangeable allusions and literary self-designations.

The reference to 'raising the dead' solves another knotty problem that much exercised Qumran commentators, namely whether those responsible for these documents held a belief in the resurrection of the dead. Though there are numerous references to 'Glory' and splendid imagery relating to Radiance and Light pervading the Heavenly abode in many texts, this is the first definitive reference to resurrection in the corpus. It should not come as a surprise, as the belief seems to have been a fixture of the Maccabean Uprising as reflected in 2 Macc. 12:44-45 and Dan. 12:2, growing in strength as it came down to first-century groups claiming descent from these archetypical events.

Text

Fragment 1

Column 2 (1)[... The Hea]vens and the earth will obey His Messiah, (2) [... and all th]at is in them. He will not turn aside from the Commandments of the Holy Ones. (3) Take strength in His service, (you) who seek the Lord. (4) Shall you not find the Lord in this, all you who wait patiently in your hearts? (5) For the Lord will visit the Pious Ones (Hassidim) and the Righteous (Zaddikim) will He call by name. (6) Over the Meek will His Spirit hover, and the Faithful will He restore by His power. (7) He shall glorify the Pious Ones (Hassidim) on the Throne of the Eternal Kingdom. (8) He shall release the captives, make the blind see, raise up the do[wntrodden.] (9) For[ev]er will I cling [to Him ...], and [I will trust] in His Piety (Hesed, also 'Grace'), (10) and [His] Goo[dness...] of Holiness will not delay ...(11) And as for the wonders that are not the work of the Lord, when He ... (12) then He will heal the sick, resurrect the dead, and to the Meek announce glad tidings. (13)... He will lead the [Holly Ones; He will shepherd [th]em; He will do (14)...and all of it... Fragment 1 Column 3 (1) and the Law will be pursued. I will free them ... (2) Among men the fathers are honored above the sons ...(3)I will sing (?)the blessing of the Lord with his favor...(4) The l[an]d went into exile (possibly, 'rejoiced) every-wh[ere...] (5) And all Israel in exil[e (possibly 'rejoicing') ...] (6) ... (7) ...

Fragment 2

(1) ... their inheritan[ce...] (2) from him ...

Fragment 3 Column 1 (4) ... he will not serve these people (5) ... strength () ... they will be great Fragment 3 Column 2 (1) And... (3) And ... (5) And ... (6) And which ... (7) They gathered the noble[s...] (8) And the eastern parts of the heavens ... (9) [And] to all yo[ur] fathers ... Fragment 4 (5) ... they will shine (6)... a man (7) ... Jacob (8)... and all of His Holy implements (9)... and all her anointed ones (10)... the Lord will speak... (11) the Lord in [his] might (12)... the eyes of Fragment 5 (1)... they [will] see all... (2) and everything in it... (3) and all the fountains of water, and the canals... (4) and those who make... for the sons of Ad[am...] (5) among these curs[ed ones.] And what ...(6) the soothsayers of my people ... (7) for you ... the Lord ... (8) and He opened...

The Messianic Leader (NASI - 4Q285)

We released this text at the height of the controversy over access to the Dead Sea Scrolls in November 1991. Since then much discussion has occurred concerning it. Our purpose in releasing it was to show that there were very interesting materials in the unpublished corpus which for some reason had not been made public and to show how close the scriptural contexts in which the movement or community responsible for this text and early Christianity were operating really were. However one reconstructs or translates this text, it is potentially very explosive.

The reader should appreciate that the Nasi ha-'Edah does not necessarily represent a Messiah per se, though he is being discussed in this text in terms of Messianic proof texts and allusions. 'Nasi' is a term used also in Column v. 1 of the Damascus Document when alluding to the successors of David. In fact, the term 'Nasi ha—'Edah' itself actually appears in CD's critical interpretation of the 'Star Prophecy' in Column vii, which follows. In its exegesis CD ties it to 'the Sceptre' as we shall see in Chapter 3 below.

There can be no mistaking this thrust in the present document, nor the parallel 4QpIs{a}. Its nationalistic thrust should be clear, as should its Messianism. If these fragments do relate to the War Scroll, then they simply reinforce the Messianic passages of the last named document. The 'Kittim' in the War Scroll have been interpreted by most people to refer to the Romans. The references to Michael and the 'Kittim' in the additional fragments grouped with the present text simply reinforce these connections, increasing the sense of the Messianic nationalism of the Herodian period. However these things may be, the significance of all these allusions coming together in a little fragment such as this cannot be underestimated.

Text

Fragment 1

(1) ... the Levites, and ha[lf...] (2) [the ra]m's horn, to blow on them ... (2) the Kittim, and

...

Fragment 2

(1)... and against ...(2) for the sake of Your Name ... (3) Michael ...(4) with the Elec[t...] Fragment 3 (2)...rain ...and spring [rain ...] (3) as great as a mountain. And the earth ... (4) to those without sense ... (5) he will not gaze with Understand[ing ...] (6)from the earth. And noth[ing ...] (7) His Holiness. It will be called ... (8) your...and in your midst...

Fragment 4

(1) ... until ... (2) you to (or '*God*')...(3)and in Heaven ...(4) in its time, and to...(5)[he]art, to...(6)and not... (7) all... (8) for *God*...

Fragment 5

(1) ...from the midst of[the]community ...(2) Riches [and] booty ... . (3) and your food . ..
(4) for them, grave[s . ..] (5) the[ir] slain... (6) of iniquity will return... (7) in compassion and...
(8) Is[r]ael...

Fragment 6

(1)... Wickedness will be smitten ...(2) [the Lea]der of the Community and all Isra[el...] (4) upon the mountains of... (5) [the] Kittim... (<) [the Lea]der of the Community as far as the [Great] Sea... (7) before Israel in that time . .. (8) he will stand against them, and they will muster against them ...(9) they will return to the dry land in th[at] time ...(10) they will bring him before the Leader of [the Community ...]

Fragment 7

(1) ... Isaiah the Prophet, ['The thickets of the forest] will be fell[ed with an axe] (2) [and Lebanon shall f]all [by a mighty one.] A staff shall rise from the root of Jesse, [and a Planting from his roots will bear fruit.'] (3)... the Branch of David. They will enter into judgement with ...(4) and they will put to death the Leader of the Community, the Bran[ch of David] (this might also be read, depending on the context, 'and the Leader of the Community, the Bran[ch of David'], will put him to death) ... (5) and with woundings, and the (high) priest will command ... (G)

[the sl]ai[n of the] Kitti[m]...

The Servants of Darkness (4Q471)

This is a text of extreme significance and another one related to the War Scroll. The violence, xenophobia, passionate nationalism and concern for Righteousness and the Judgements of *God* are evident throughout. Though these may have a metaphoric meaning as well as an actual one, it is impossible to think that those writing these texts were not steeped in the ethos of a militant army of *God*, and hardly that of a peaceful, retiring community. Their spirit is unbending, uncompromising. They give no quarter and expect none.

One should also note, in particular, the widespread vocabulary of 'Judgement', the 'Heavenly Hosts' and even 'pollution'. Notice, too, the consistent emphasis on 'Righteousness' and 'Righteous judgement', and on 'keeping', i.e. 'keeping the Law' - 'Covenant' in this text. The group responsible for these writings is extremely Law oriented and their zeal in this regard is unbending. The very use of the word 'zeal' connects the literature with the Zealot mentality and movement.

Text

Fragment 1

(1) ... the time You have commanded them not to (2) ... and you shall lie about His Covenant (3) ... they say, 'Let us fight His wars, for we have polluted (4)... your [enemie]s shall be brought low, and they shall not know that by fire (5)... gather courage for war, and you shall be reckoned (6)... you shall ask of the experts of Righteous judgement and the service of (7)... you shall be lifted up, for He chose [you]... for shouting (8) ... and you shall bur[n...]and sweet...

Fragment 2

(2) to keep the testimonies of our Covenant ...(3) all their hosts in forbear[ante...] (4) and to restrain their heart from every w[ork ...(5) Se]wants of Darkness, because the judgement ...(6) in the guilt of his lot... (7) [to reject the Go]od and to choose the Evil... (8) *God* hates and He will erect ... (9) all the Good that...

Fragment 3

(2) Eternal, and He will set us ...(3)[He jud]ges His people in Righteousness and [His] na[tion in ...] (4) in all the Laws of ...(5) us in [our]sins...

Fragment 4

(1) from all tha[t...] (2) every man from his brother, because (3) ... and they shall remain with Him always and shall se[rvel (4) ... each and every tribe, a man (5) . .

[twen]ty-[six] and from [the] Levites six(6) [teen...] and [they] shall se[rve before Him] always upon (7)... [in] order that they may be instructed in ...

The Birth of Noah (4Q534-536)

A pseudepigraphic text with visionary and mystical import, the several fragments of this text give us a wonderfully enriched picture of the figure of Noah, as seen by those who created this literature. In the first place, the text describes the birth of Noah as taking place at night, and specifies his weight. It describes him as 'sleeping until the division of the day', probably implying noon. In this text, too, the Kabbalistic undercurrents should be clear and the portrayal of Noah as a Wisdom figure, or one who understands the Secret Mysteries, becomes by the end of Fragment 2 its main

Text

Fragment 1

(1) ... (When) he is born, they shall all be darkened together... (2) he is born in the night and he comes out Perfe[ct...] (3) [with] a weight of three hundred and fifty] shekels (about 7 pounds, 3 ounces)... (4) he slept until the division of the days... (5) in the daytime until the completion of years... (6) a share is set aside for him, not ...years...

Fragment 2

Column 1 (1)... will be ...(2) [H]oly Ones will remem[ber ...] (3) lig[hts] will be revealed to him (4)... they [will] teach him everything that (5)... human [Wi]sdom, and every wise ma[n...] (6) in the lands(?),and he shall be great (7)... mankind shall[be]shaken, and until (8)... he will reveal Mysteries like the Highest Angels (9)... and with the Understanding of the Mysteries of (10)... and also (11) ... in the dust (12) ... the Mystery [as]cends (13) ... portions ...

Fragment 3

Column 2 (7) from... (8) he did... (9) of which you are afraid for all... (1 0) his clothing at the end in your warehouses. (?) I will strengthen his Goodness ...(11) and he will not die in the days of Wickedness, and the Wisdom of Your mouth will go forth. He who opposes You (12) will deserve death. One will write the words of *God* in a book that does not wear out, but my words (13) you will adorn. At the time of the Wicked, me will know you forever, a man of your servants...

Fragment 4

Column 1(1) ... of the hand, two ... it lef[t] a mark from ... (2) barley [and] lentils on ... (3) and tiny marks on his thigh ...[Aftertw]o years he will be able to discern one thing from another ... (4) In his youth he will be... all of them ...[like a ma]n who does not know anyth[ing, until] the time when (5) he shall have come to know the Three Books. (6) [Th]en he will become wise and will be disc[rete ...] a vision will come to him while upon [his] knees (in prayer). (7) And with his father and his forefa[th]ers... life and old age; he will acquire counsel and prudence, (8)[and] he will know the Secrets of mankind. His Understanding will spread to all peoples, and he will know the Secrets of all living things. (9)[A1]1 their plans against him will be fruitless, and the spiritual legacy for all the living will be enriched. (10) [And all] his [p]lans [will succeed], because he is the Elect of *God*. His birth and the Spirit of his breath (11) ... his[p] fans will endure forever ... (12) that ... (13) pl[an ...

Words of Michael (4Qmich, 6Qunidar, 4Q529)

The Archangel Michael is the "protector of Israel"; he plays a prominent role in Jewish literature. Michael is considered as the leader of the angels and for this reason is considered the Chief messenger of God. In The Words of the Archangel Michael, he is portrayed as talking to other angels. Also he is seen in the text as being a given a vision from Gabriel. It is this vision that causes speculation of to what city they are actually referring to within the text. It at first looks like the author is starting to talk about the tower of Babel, but after further consideration it looks as if he is talking about Jerusalem and the building of the temple. If this assumption is right, then Michael may be asking why there is an angelic force stationed on the mountains. Gabriel vision may be an explanation to why they are there and a premonition to the great city that is to be built there.

Text

(1) The words of the book that Michael spoke to the Angels of *God* [after he had ascended to the Highest Heaven.] (2) He said 'I found troops of fire there .. . (3) [Behold,] there were nine mountains, two to the eas[t and two to the north and two to the west and two] (4)[to the south. There I beheld Gabriel the Angel... I said to him, (5)'... and you rendered the vision comprehensible.' Then he said to me . .. (6) It is written in my book that the Great One, the Eternal Lord... (7) the sons of Ham to the sons of Shem. Now behold, the Great One, the Eternal Lord ...(8) when ... tears from... (9) Now behold, a city will be built for the Name of the Great One, [the Eternal Lord]... [And no] (10) evil shall be committed in the presence of the Great One, [the Eternal] Lord ...(11) Then the Great One, the Eternal Lord, will remember His creation [for the purpose of Good]... [Blessing and honor and praise](12)[be to] the Great One, the Eternal Lord. To Him belongs Mercy and to

Him belongs... (13) In distant territories there will be a man ...(14) he is, and He will say to him, 'Behold this... (15) to Me silver and gold ...'

The New Jerusalem (4Q554, 2QExc, 4QJMa, 5QJNar, 11QJN)

It was written in Aramaic and paralleled Ezekiel xl - xliii, as well as, Revelation xxi. It is thought that a surveyor of the era was a visionary in ancient Judea who provided a detailed picture of the entire city's dimensions. The parallels this work has to the Hebrew Bible is that both measure the city in detail from east to west escorted by the guidance from a heavenly being. Ezekiel was a prophet that earnestly awaited the restoration of Israel to its once prosperous state. Other parallels like Isaiah and the book of Tobit speak of a rejuvenated city and temple of the Lord. Revelation is less detailed in the actual measuring, but more vivid in depicting the visual heavenliness of the city with references to jewels, gold, and a running crystal stream. Ezekiel and Revelation cover more of the rules governing the Lord's people and the manner in which the twelve tribes of Israel should divide the city. The purpose to the Qumran community is that of a basic picture of a rewarding place for following the laws of the Lord. Judeaism and Christianity are similar in many basic concepts in their respective scriptures.

4Q554

Frag. 1 Col. 1
[...]he measured 35 stadia from north to the southern corner and named the gate the gate of Simeon.

[From this gate he measured 35 stadia to] the middle gate which was called the gate of Levi.

From this gate he measured 35 stadia to the south which was called the gate of Judah.

From this gate he measured to the [southeastern] corner and then westwards 35 stadia and called this gate the gate of Joseph.

[He measured] 24 stadia from here to the middle and called the gate the gate of Benjamin.

From here he measure 24 stadia to the [third] gate and called it the gate of Reuben.

From here [to the western corner he measured 24 stadia] and then **Col. 2** [northwards] 35 stadia and called this gate the gate of Issachar.

He measured 24 statia from this gate to the middle and named it the gate of Zebulun.

From here he measured 24 stadia to the third gate and called this gate the gate of Gad.

From here he measured to the northern corner 35 stadia and then eastwards 35 stadia calling this gate the gate of Dan.

He measured from here to the middle 24 stadia and called this gate the gate of Naphtali.

From here he to the third gate 24 stadia and called this gate the gate of Asher.

He measured from here to the eastern corner 24 stadia.

Then he took me into the city to measure all the city blocks. He measured the length and width of the blocks to be a 51 x 51 rod square **[4Q554 + 5Q15, Frag. 1 col. I]** (357 cu. on each side). The portico of the street measured 3 rods (21 cu.). He showed me all the measurements of all the blocks. Each street between the blocks measuring 6 rods in width (42 cu.). Two main streets running East to West measured 10 rods (70 cu.) in width with the third street (which runs by the left of the temple) measuring 18 (126 cu.). The two streets running South to North measured 9 rods, 4 cu. in width (67 cu.) with the main one in the middle he measured at 13 rods, 1 cu. (92 cu.). All the city streets are paved of white stone, alabaster and onyx. [vacat]

The [...] eighty posterns were then measured: each 2 rods (14 cu.) with stone jambs measuring 1 rod (7 cu.). He showed me the dimension of the twelve [gates]. Their doors' widths were 3 rods (21 cu.). Each door had two jambs measuring 1½ rods (10½ cu.). On either side of each of the doors were towers. Their height and width were 5 rods by 5 (35 cu.). A staircase runs by the inner door, going up to the height of the towers being 5 cu. wide. The towers and the staircases are each 5 rods, 5 cu. square (40 cu. on each side of the door)[....] He showed me that the porches of the blocks were 2 rods (14 cu.) in width, and the width of the [...] measured in cubits. He measured the top of each threshold with its jambs, measuring inside 13 (length) by 10 cu. (width). He then led me inside the vestibule where there was another threshold and door on the right side of the inner wall. The wall was proportional to the outer gate, and measured 4 cu. wide by 7 cu. high. He measured the door to the room, measuring 1 rod in width. **Col. II** (7 cu.). The length of the entrance was 2 rods (14 cu.), with a height of 2 rods (14 cu.). The corresponding door had the same dimensions as they left the room. To the left he showed me a stairwell that goes around and up, with identical dimensions, 2 rods by 2 (14 cu.). The doors opposite are the same size. A pillar stands in the middle of the staircase that goes up and around which measures 6 by 6 cu. **[5Q15 + 4Q555]** The staircase, which goes up beside it, measures 4 cu. wide and ascends 2 rods up to [....]

He brought me to the interior of the city block and showed me the houses between the gates, fifteen in all. Eight went one direction to the corner gate and seven in another direction to the other gate. The houses were 3 rods (21 cu.) long by 2 rods (14 cu.) wide. They all have the same floor plan, and they are each 2 rods (14 cu.) high. Each has a 2 rod (14 cubit) door in the middle of the house. He measured the interiors of the houses[... ? An interior feature was ?...] 4 cu. in length and 1 rod (7 cu.) high. The site has 19 cu. long and 12 wide. The house has 22 beds, and eleven lattice windows above [...]. On the side was an outer gutter[...] the window, 2 cu. high [...] thickness and width of the wall [...] the platform, 19 cu. wide [and 12] cu. wide. [...] their height [...] 2 rods (14 cu.) [... a width] of 3 cu. and a length of 10 [cu....] 1½ cu.[...]

4Q554

Frag. 2 Col. 2
[...]its foundation. It was 2 rods (14 cu.) wide and 7 (49 cu.) high. All of it built of electrum and sapphire and chalcedony with beams of gold. It had 1432 towers whose length equated their width and with heights of 10 rods (70 cu.).

[the text continues with a description of the sacrificial activities in the new temple and a prophecy about the surrounding nations]

The Tree of Evil (A Fragmentary Apocalypse-4Q458)

This Hebrew apocalypse, while fragmentary, again recapitulates themes known across the broad expanse of Qumran literature, most notably tem{c}a (polluted), teval{c}a (swallow or swallowing), 'walking according to the Laws', yizdaku (justified or made Righteous), etc. These themes should not be underestimated and reappear repeatedly in the Damascus Document, the Temple Scroll, Hymns, and the like. As fragmentary as the Tree of Evil text is, there are apocalyptic references to 'Angels', 'burning', 'flames', etc. Images like 'burning fire' have an almost Koranic ring to them, as do references to 'the moon and stars'. There is also an intriguing reference to 'the beloved one' - possibly referring to Abraham as 'friend of *God*' - of the kind one finds in texts like the Damascus Document and notably the Letter of James.

Text

Fragment 1

(1) ... to the beloved one ... (2) the beloved one ... (3) in the tent... (4) they did not know... (5) burning of fire... (6) and the peoples of the ... arose ... (7) spoke to the first, saying ...(8) flames, and He will send the first Angel... (9) drying up. And he smote the Tree of Evil...

Fragment 2 Column 1 (2) [... the mo]on and the stars (3)... the years (4)... he fled in (5) ... the polluted (one) (6) ... the harlots(?)

Fragment 2 Column 2 (2) And he destroyed him, and ...(3) and swallowed up all the uncircumcised, and it ...(4) And they were justified, and walked according to the L[aws... (5) anointed with the oil of the Kingship of ...

Fernando Klein

Vision of Jacob (4QAJa=4Q537)

The first fragment is the continuity of Jacob's first vision described in Genesis 28:10-19[3] in the Bible after he set up a stone and poured out a libation upon it. In this second vision, God confirms his Covenant with Jacob by promising him blessing and righteous. In return, Jacob accepts God as the only God. He also voluntarily adds two conditions to the agreement. First, he promises that he will give one tenth of what he earns back to God. Second, he affirms that the stone, which he established, will serve as the foundation for a sanctuary to God, to be built upon his return. Fragment 1 also foretells that Bethel was not the place God ultimately chose for his Temple, which indicated in the extrabiblical book Jubilees.

In fragment 2, the text reveals an eschatological figure of the High Priest of the messianic era who makes the expiation for the people. His mission is to be a suffering servant to encounter human's sins. In order to do so, the priest need to suffer, die, (or even be crucified). All of these are alluded in the text, however, they are supposed to have been made in the end of the second century BCE.

4Q537

Frag.1

[Then I had a vision at night. An angel of God came down from heaven with seven tablets in his hand. He told me, "God Most High has blessed you, and] $_1$ your later generations. All just and upright men will survive [...and no more] $_2$ evil [will be done]; lying should not be found among [...] $_3$ Now, take the tablets and read everything [that is written on them." So I took the tablets and read. There were written all my sufferings,] $_4$ troubles and everything that would happen to me [during the one hundred and forty seven] years of my life. [Then he told me," Take] this tablet." [...] $_5$ [So] I took that tablet [and ... read everything on it.] I saw that it said [no temple should be built in this place,] $_6$ [... Then he told me,] "you would leave here on the [eighth] day [... and your offerings would not be] invalid before [God Most High..."] $_7$ [...] ... [...]

Frag. 2

$_1$ [I saw...] and how will the building be built [... how] priests will be dressed, and [their hands] be purified, $_2$ [and how] they will offer sacrifices on the altar. And how they will eat part of their sacrifices [on the who]le earth $_3$ [...and drink the water] that will come from the city beneath the walls, and where they [...] $_4$ [...] Blank [...] $_5$ [...Then I looked,] before me was a land divided into two squares and [...]

Midrash on Last Days (4Q174, 4QFlorilegium)

This text is what remains of a collection of Old Testament texts considered messianically and eschatologically significant along with some commentary. The author interprets an abbreviated version of 2 Sam 7:11c-14a as messianic, on the assumption that God is referring not to Solomon but to David's greatest "son" or descendent, the eschatological Davidic king.

In his commentary on this passage, the author explicitly identifies the "son" in 2 Sam 7:11c-14a as the "the branch of David." This means that the author has identified David's "son" in 2 Sam 7:14 with the eschatological Davidic king described metaphorically as the "branch of David" in Jer 23:5; 33:15. In 4Q174 1.12b, Amos 9:11 is quoted as referring to the appearance of this Davidic king: ""I will raise up the tent of David that has falle[n] (Amos 9:11), who will arise to save Israel." (1.13). He is destined to "save Israel" (lhwšy` 'th yšr'l), by which doubt is meant a political and military deliverance. Similarly, in 4Q174 1.18, Ps 2:1 is quoted and interpreted: 18 "[Why] do the nations [rag]e and the people im[agine] a vain thing? [Kings of the earth] ris[e up] and [and p]rinces conspire together against Yahweh and against [his anointed] (Ps 2:1-2). 19 [In]terpretation of the saying [concerns na]tions and th[ey   ] the chosen of Israel in the last days. Although the text is not complete, it is clear that Ps 2:1-2 is being interpreted messianically. The anointed one, against whom the nations rage, is called the "elect of Israel in the last days," meaning the eschatological Davidic king.

4QFlorilegium

Col. I (Frgs. 1-3)

1 . . . [I will appoint a place for my people Israel and will plant them in order that they may dwell there and no more be troubled by their] enemies. No son of iniquity [will afflict them again] as before, from the day that 2 [I set judges] over my people Israel (2 Sam 7:10). This is the house which [in the] last days according as it is written in the book 3 [the sanctuary, O Lord,] which your hands have established, Yahweh shall reign for ever and ever (Exod 15:17-18) This is the house in which [ ] shall not enter there 4 [f]orever, nor the Ammonite, the Moabite, nor the bastard, nor the foreigner, nor the stranger forever because there shall be the ones who bear the holy name 5 [f]orever. Continually it will appear above it. And strangers will no longer destroy it as they previously destroyed 6 the sanctuary of Israel because of its sins. He commanded that a sanctuary of men be built for himself in order to offer up to him like the smoke of incense 7 the works of the Law And according to his words to David, (2 "And I [will give] you [rest] from all your enemies" (2 Sam 7:11). This means that he will give them rest from a[ll] 8 the sons of Belial, who cause them to stumble to destroy them [ ] according as they come with a plan of [B]el[i]al to cause the s[ons of] 9 light to stumble, to think upon them wicked plans in order to deli[ver] his [s]oul to Belial in their w[ic]ked error. 10 [And] Yahweh has [de]clared to you that he will build you a house (2 Sam 7:11c). I will raise up your seed after

you (2 Sam 7:12). I will establish the throne of his kingdom 11 f[orever] (2 Sam 7:13). I wi[ll be] a father to me and he shall be a son to me (2 Sam 7:14). He is the branch of David who will arise with the interpreter of the Law who 12 [ ] in Zi[on in the la]st days according as it is written: "I will raise up the tent of 13 David that has falle[n] (Amos 9:11), who will arise to save Israel. 14 An in[ter]pretation of "Blessed is [the] man who does not walk in the counsel of the wicked" (Ps 1:1). Interpretation of the wor[d concerns] those who depart from the way [ ] 15 which is written in the Book of Isaiah the prophet for the last [d]ays, "It happened that with a strong [hand he turned me aside from walking on the path] of 16 this people" (Isa 8:11). And they are those about whom it is written in the Book of Ezekiel the prophet, "[They should] not [defile themselves any longer with all] 17 their idols (Ezek 37:23; see 44:10). These are the sons of Zadok and the m[e]n of his his cou[ns]el [ ] after them to the council of the community. 18 "[Why] do the nations [rag]e and the people im[agine] a vain thing? [Kings of the earth] ris[e up] and [and p]rinces conspire together against Yahweh and against [his anointed] (Ps 2:1-2). 19 [In]terpretation of the saying [concerns na]tions and th[ey] the chosen of Israel in the last days.

Col. 2 (Frgs. 1-3)

1. This is the time of the trial that c[omes J]udah to complete [ ] 2 Belial, and a remnant will remain [l]ot and they do all the Law [ ] 3 Moses. It is [a]s it is written in the Book of Daniel, "The wicked [act wickedly]" 4a and the righteous [shall be made wh]ite and be purified (Dan 12:10) And a people who know God will remain strong [ ] . After [ ] which is for them [ ] in their descent

Prophecy and Apocalyptic

Fernando Klein

The Chosen One (4Qelect, 4QarNC)

The fact that all things happen according to God's divine plan for the world is a fairly common theme among the Dead Sea Scrolls. For example, it was thought that if you belonged to the Qumran sect, then it was so because it was in God's plan that you belonged to it. For this reason, the Qumran sect sometimes referred to themselves as "the chosen of God." However, the following texts refer to a particular person as the "chosen one." It was originally thought that the "chosen one" referred to a messiah (if not the messiah). In 4QTLevi (4Q541) there is a prophecy of an eschatological priest reminiscent of the man described in this text (Wise, 428). However, others believe that it is just as likely that the text alludes to the miraculous birth of Noah. For this reason, it is sometimes placed together with the remains of other Noah literature. Vermes believes that this with a few other Qumran fragments "appear to be the relics of a Book of Noah mentioned in Jubilees x, 13 and xxi, 10." (521).

The scroll marked 4Q534 is also labeled under the designations "4Qmess ar" and "4QElect of God". This scroll tells what the "Chosen One" will look like and some about his education and future greatness. The scroll 4Q535 (also designated "4QAramaic N") tells more about the circumstances of his birth, although details are obscure. The last of the scrolls in this set, 4Q536 (also designated "4QAramaic C") tells of the "Chosen One's" teachings.

4Q534

Col 1

$_1$[...] of his hand, two [...] a mark. His ₁ hair will be red and he will have moles on [...] ₁ and small marks in his thighs. [And after t]wo years, he will know one thing from another. ₁ While he is young, he will be like ...[...like] someone who knows nothing, until he $_5$knows the three Books [...] ₁ Then he will gain wisdom and learn understanding [...] visions will come to him while he is on his knees. ₁ And with his father and ancestors [...] life and old age. He will have wisdom and discretion ₁ and he will know the secrets of man. His wisdom will reach out to everyone and he will know the secrets of all living things. ₁ All of their plans against him will fail, and his rule over all things will be great. $_{10}$[...] his plans will succeed because he is the one picked by God. His birth and the breath of his spirit [...] and his plans will last forever. [...]

Col 2

$_1$[...] which [...] fell in ancient times. The sons of the pit [...] ₁ [...] evil. The spot [...] ₁ [...] ₁ [...] in order to go [...] $_5$ [...] flesh [...] ₁ [...] ₁ and his breathing out [...] ₁ forever [...] ₁ $_{10}$ ₁ [...] ₁ and the cities [...] ₁ and they will destroy [...] ₁ The waters will stop [...] they will destroy [...] from the heights. They will all come [...] $_{15}$ [...] ₁ [...] and they will all be destroyed. His work will be like that of the Watcher. ₁ Instead of his voice [...] he will establish his foundation on him. His sin and his error ₁ [...] the Holy One and the Watchers [...] to say ₁ they will speak against him [...].

The Dead Sea Jesus

4Q435

frag. 1
₁when [...] ₁ Baraq'el [...] ₁ my face once more [...] ₁ I got up [...]
frag. 2
₁[...]the time of birth [...] ₁ [...] the walls of the house of [...]
frag 3
₁[...] he is born and they are praised together [...] ₁ [...] he is born at night and comes out complete [...] ₁ [...] with the weight of three hundred and fifty shekels (a Hebrew unit equal to about 252 grains troy) [...] ₁ [...] he sleep until mid afternoon and [...] ₅ [...] during the day until two years are over [...] ₁ [...] he removes it from him; and after [x] years [...]

4Q536

frag 1 Col 1
₁[...] you will be [...] ₁ [...] he will make you think of the holy angels [...] ₁ [...] the lights will be revealed to him ₁ [...] all of his teachings ₅ [...] the wisdom of humanity, and every wise man ₁ [...] in the region he will be great ₁ [...] humanity will be troubled ₁ [...] he will share God's secrets ₁ [...] he will understand God's mysteries [...]
frag 1 Col 2
₈ he made [...] ₁ that you are afraid of [...] ₁₀ he will strengthen its concealment at the end of your powers. His possessions [...] ₁ and he will not die in the days of evil. And his words will contain great wisdom. I will praise you [...] ₁ is sentenced to death. Who will write the words of God in a book that will not decay? And my sayings [...] ₁You will come to me and in the time of evil he will know you forever. A man who [...] your servants, [...] sons [...]

The Book of Secrets (1Q27, 4Q299-301)

4Q301 F1
(...) I shall speak out freely, and I shall express my various sayings among you (...) (.. those who would understand parables and riddles, and those who would penetrate the origins of knowledge, along with those who hold fast to the wonderful mysteries ...) (...) those who walk in simplicity as well as those who are devious in every activity of the deeds of humanity ...) those with a stiff neck, a hard pate, and all the mass of the Gentiles, with (...)

4Q301 F2
the customs of the fool and the inheritance of the wise (...) Now what good is the riddle to you, you who search for the origins of knowledge? Why is the heart

honoured, for it is the dominion (...) a parable? Why is it splendid to you, for it is (...) Why is a prince (...) ruler? (...) without strength, and he dominates him with a whip that cost nothing. Who could say (...) who among you seeks the presence of Light and Illumination (...) the plan of memory without (...) (...) by the angels of (...) (...) those who praise (...)

4Q300 F3
so that they would know the difference between good and evil .....)

1Q27 col 1
secrets of sin (...) but they did not know the secret of the way things are nor did they understand the things of old and they did not know what would come upon them, so they did not rescue themselves without the secret of the way things are. This shall be the sign that this shall come to pass : when the sources of evil are shut up and wickedness is banished in the presence of righteousness, as darkness in the presence of light, or as smoke vanishes and is no more, in the same way wickedness will vanish forever and righteousness will be manifest like the sun. The world will be made firm and all the adherents of the secrets of sin shall be no more. True knowledge shall fill the world and there will never be any more folly. This is all ready to happen, it is a true oracle, and by this it shall be known to you that it cannot be averted.
It is true that all the peoples reject evil, yet it advances in all of them. It is true that truth is esteemed in the utterances of all the nations - yet is there any tongue or language that grasp it? What nation wants to be oppressed by another that is stronger? Or who wants his money to be stolen by a wicked man? Yet what nation is there that has not oppressed its neighbour? Where is the people that has not robbed the wealth of another ...

4Q299 F2 (+ 4Q300 F5) Col 2
what should we call a man who ... his) deeds (...) but every deed of the righteous has been judged impure. And what shall we call man who (... call no one on earth) wise or righteous, for it is not a human possession (...) and not (...) (...wisdom is hidden) except for the wisdom of cunning evil, and the schemes of Belial ...) a thing that ought never to be done again, except (...) the command of his Maker ; and what shall a man do and live? ... he who) has violated the command of his Maker shall have his name erased from the mouth of all (...) (...) So listen, you who hold fast to the wonderful secrets ...) of eternity , and the plots behind every did, and the purpose of .... He knows) every secret and stands behind very thought. He does every (... the Lord of all) is He, from long ago He established it, and forever (...) (...) the purpose of the origins he opened up to (...) (...) for he tests His son, and gives as an inheritance (...) (...) every secret, and he limits of every deed; and what (...) (...) the Gentiles, for He created them and their deeds (...)

4Q300 F1 Col 2
Consider the soothsayers, those teachers of sin. Say the parable, declare the riddle before we speak ; then you will know if you have understood. (...) your foolishness,

for the vision is sealed up from you, and you have not properly understood the eternal mysteries and you have not become wise in understanding (...) (...) for you have not properly understood the origin of Wisdom; but if you should unseal the vision (...) (...) all your wisdom, for to you (...) Hear now what wisdom is.

4Q299 F5
(...light)s of the stars for a memorial of His name ...)(...hidden) things of the mysteries of Light and the ways of Darkness (...) (...) the times of heat with the periods (of cold....) (... the breaking of day) and the coming of night (...) (...) the origins of things (...).

4Q299 F8
(...) How can a man understand without knowledge or hearing? (...) (...) He created insight for His children, by much wisdom He uncovered our ears tat we may h(ear...) (...) He created insight for all those who pursue true knowledge and (...) (...) all wisdom is from eternity; it may not be changed (...) (...) He locked up behind the waters, so that not (...) (...) the heaven above heaven (...)

4Q301 F3
(...) and He is well known for His patience, and might in His great anger, and splendid (...) He in His numerous acts of mercy, and terrible in His wrathful purposes, and honoured (...) (...) and over the land He made him a ruler, and God is honoured among His Holy people, and splendid among His chosen, yes, splendid (...) holy, great in the blessing of (...) (...) their splendour and (...) when the Era of Wickedness is at an end, and evil doing (...)

The Divine Throne Chariot

The Divine Throne-Chariot draws its inspiration from Ezekiel (1:10) and is related to the Book of Revelation (4). It depicts the appearance and movement of the Merkabah, the divine Chariot supported and drawn by the cherubim, which is at the same time a throne and a vehicle. The "small voice" of blessing is drawn from 1Kings 19:12: it was in a "still small voice" that God manifested himself to Elijah. In our Qumran text this voice is uttered by the cherubim and it is interesting to note that although the Bible does not define the source of the voice, the ancient Aramaic Text of 1Kings (Targum of Jonathan) ascribes it to angelic beings called "they who bless silently." The Throne-Chariot was a central subject of meditation in ancient as well as in medieval Jewish esotericism and mysticism, but the guardians of Rabbinic orthodoxy tended to discourage such speculation. The liturgical use of Ezekiel's chapter on the Chariot is expressly forbidden in the Mishnah; it even lays down that no wise man is to share his understanding of the Merkabah with a person less enlightened than himself. As a result, there is very little ancient literary material extant on the subject, and the Qumran text is therefore of great importance to the study of the origins of Jewish mysticism.

"...The ministers of the Glorious Face in the abode of the gods of knowledge fall down before him, and the cherubim utter blessings. And as they rise up, there is a divine small voice and loud praise ; there is a divine small voice as they fold their wings. The cherubim bless the image of the Throne-Chariot above the firmament, and they praise the majesty of the fiery firmament beneath the seat of his glory. And between the turning wheels, angels of holiness come and go, as it were a fiery vision of most holy spirits; and about them flow seeming rivulets of fire, like gleaming bronze, a radiance of many gorgeous colors, of marvelous pigments magnificently mingled. The Spirits of the Living God move perpetually with the glory of the wonderful Chariot. The small voice of blessing accompanies the tumult as they depart, and on the path of their return they worship the Holy One, Ascending they rise marvelously; settling, they stay still. The sound of joyful praise is silenced and there is a small voice of blessing in all the camp of God. And a voice of praise resounds from the midsts of all their divisions in worship. And each one in his place, all their numbered ones sing hymns of praise."

The Coming of Melchizedek (11Q13)

Col.2

(...) And concerning what Scripture says, "*In this year of Jubilee you shall return, everyone f you, to your property"* (**Lev. 25;13**) And what is also written; "*And this is the manner of the remission; every creditor shall remit the claim that is held against a neighbor, not exacting it of a neighbor who is a member of the community, because God's remission has been proclaimed"* (**Deut.15;2**) the interpretation is that it applies to the Last Days and concerns the captives, just as Isaiah said: "*To proclaim the Jubilee to the captives"* (**Isa. 61;1**) (...) just as (...) and from the inheritance of Melchizedek, for (... Melchizedek) , who will return them to what is rightfully theirs. He will proclaim to them the Jubilee, thereby releasing them from the debt of all their sins. He shall proclaim this decree in the first week of the jubilee period that follows nine jubilee periods.

Then the "*Day of Atonement"* shall follow after the tenth jubilee period, when he shall atone for all the Sons of Light, and the people who are predestined to Melchizedek. (...) upon them (...) For this is the time decreed for the "*Year of Melchizedek`s favor"*, and by his might he will judge God's holy ones and so establish a righteous kingdom, as it is written about him in the Songs of David ; "*A godlike being has taken his place in the council of God; in the midst of divine beings he holds judgement"*

(**Ps. 82;1**). Scripture also says about him ; "*Over it take your seat in the highest heaven; A divine being will judge the peoples"* (**Ps. 7;7-8**) Concerning what scripture says; "*How long will you judge unjustly, and show partiality with the wicked? Selah"* (**Ps. 82;2**) ,the interpretation applies to Belial and the spirits predestined to him, because all of them have rebelled, turning from God's precepts and so becoming utterly wicked. Therefore Melchizedek will thoroughly prosecute the vengeance required by God's statutes. Also, he will deliver all the captives from the power of Belial, and from the power of all the spirits destined to him. Allied with him will be all the *"righteous divine beings"* (**Isa. 61;3**).

(The ...) is that whi(ch ...all) the divine beings. The visitation is the Day of Salvation that He has decreed through Isaiah the prophet concerning all the captives, inasmuch as Scripture says, "How beautiful upon the mountains are the feet of the messenger who announces peace, who brings good news, who announces salvation, who says to Zion "Your divine being reigns"." (**Isa. 52;7**) This scriptures interpretation : "the mountains" are the prophets, they who were sent to proclaim God's truth and to prophesy to all Israel. "The messengers" is the Anointed of the spirit, of whom Daniel spoke; "After the sixty-two weeks, an Anointed shall be cut off" (**Dan. 9;26**) The "messenger who brings good news, who announces Salvation" is the one of whom it is written; "to proclaim the year of the LORD`s favor, the day of the vengeance of our God; to comfort all who mourn" (Isa. 61;2)

This scripture's interpretation: he is to instruct them about all the periods of history for eternity (... and in the statutes) of the truth. (...) (.... dominion) that passes from Belial and returns to the Sons of Light (....) (...) by the judgment of God, just as t is written concerning him; "who says to Zion "Your divine being reigns" (Isa. 52;7) "Zion" is the congregation of all the sons of righteousness, who uphold the covenant and turn from walking in the way of the people. "Your divine being" is Melchizedek, who will deliver them from the power of Belial. Concerning what scripture says, "Then you shall have the trumpet sounded loud; in the seventh month…" (Lev. 25;9)

Messianic Apocalypse (4Q521)

The Qumran text designated as 4Q521 consists of 16-18 fragments. Fragment 2, which consists of three columns, is the most important of the fragments. This fragment appears to be a description of the conditions that will obtain at the time of eschatological salvation.

In line one of frg. 2, col. 2, there is a reference to "his anointed." The antecedent of "his" is probably God, since God, identified as lord ('dny), occurs in line three; the anointed is probably the eschatological Davidic king. The fact that the heavens and the earth obey God's anointed one means that God has put the entire universe under his authority (see Deut 32:1; Isa 1:2; see Ps 146:6). This fragmentary text refers to God's Messiah (or anointed) whom the heavens and earth will obey. Although what follows is a description of God's eschatological activities, it seems that the Messiah may be the instrument through which some or all of these will be accomplished. Several Old Testament quotations and allusions occur.

One of the more intriguing of the newly released Dead Sea Scrolls is a fragment now titled "Messianic Apocalypse" (4Q521). This text contains three rather striking features that are of particular significance for comparing the apocalyptic beliefs and expectations of the Qumran community with the emerging early Christian movement. First, the text speaks of a single Messiah figure who will rule heaven and earth. Second, it mentions in the clearest language the expectation of the resurrection of the dead during the time of this Messiah. And third, and perhaps most important for students of the New Testament, it contains an exact verbal parallel with the Gospels of Matthew and Luke for identifying of the signs of the Messiah.

4Q521

[the hea]vens and the earth will listen to His Messiah, and none therein will stray from the commandments of the holy ones. Seekers of the Lord, strengthen yourselves in His service! All you hopeful in (your) heart, will you not find the Lord in this? For the Lord will consider the pious (hasidim) and call the righteous by name. Over the poor His spirit will hover and will renew the faithful with His power. And He will glorify the pious on the throne of the eternal Kingdom. He who liberates the captives, restores sight to the blind, straightens the b[ent] And f[or] ever I will cleav[e to the h]opeful and in His mercy . . . And the fr[uit . . .] will not be delayed for anyone.

Prophets and Pseudo-Prophets

Fernando Klein

The Angels of Mastemoth and the Rule of Belial (4Q390)

Since it is written in the first person rather than the third, however, and is evidently meant to be a direct expression of *God*'s words, we place it in this prophetic section. Relating to both Ezekiel and Daniel, it contains an allusion from Hosea as well. The text, which could be referred to as a pseudo-Moses text, or even possibly a pseudo-Aaron one, also has strong thematic parallels with Jubilees and Enoch.

Its parallels with the exhortative section of the introductory columns of the Damascus Document are intrinsic, including an emphasis on 'breaking the Covenant' (CD,i.20), 'pollution of the Temple' (iv.18), 'going astray' (i.15) and 'walking in the stubbornness of their hearts' found there.(ii17, iii. 5, 11-12). The expression 'They will pollute My Temple' directly parallels what goes under the heading of one of the 'three nets of Belial' in the Damascus Document.

These are 'fornication', 'riches' and 'pollution of the Temple', which Belial is characterized as setting up 'as three kinds of Righteousness' and by which he is said to have 'taken hold of Israel'. Of course, allusion to 'the rule of Belial' is strong in both the Angels of Mastemoth and the Damascus Document texts, as it is in many of the documents noted under our comments concerning 'swallowing'. allusion above.

The chronology of this apocalypse to a certain extent follows jubilees and brings us down to the same period presaged in the Damascus Document. There is also direct reference to the 'seventy years' of Dan. 9:2. The only question is whether the chronology followed by these literary practitioners is any more exact than that encountered in Josephus or Talmudic traditions, which is often not reliable at all. Do they have a clear idea of seven jubilees in absolute chronological terms?

There is also an anti-priestly thrust to the apocalypse, in the sense that as in Ezekiel the priests have been 'warned', but their breaking of the Law and the Covenant, robbing of Riches, and violence goes even as far as 'polluting the Temple'. Whether this relates to a pre-Maccabean, the Maccabean, or the Herodian period is difficult to say, but the unbending, nationalist and anti-corruption stance is constant. Nor is this stance particularly retiring or uninterested in the affairs of men.

Text

Fragment 1

(2) [and] break[ing...] again (?)... the sons of Aaron... seventy years ...(3) and the sons of Aaron shall rule them, but they shall not walk in My Wa[ys,] which I comm[an]d you and which (4) you shall warn them about. They also (i.e., sons of Aaron) shall do what is Evil in my eyes, exactly as Israel did (5) in the early days of its 7Kingdom-apart from those who will come up first from the land where they

have been captive, to build (6) the Temple. And I will speak to them and send them Commandments, and they will understand to what extent (7) they have wandered astray, they and their forefathers. But from the end of that generation, corresponding to the Seventh jubilee (8) since the desolation of the land, they will forget Law and festival, Sabbath and Covenant. They will break (i.e., violate) everything, and do (9) what is evil in My eyes. Thus I shall turn My face away from them, and give them into the hands of their enemies, delivering [them] (10) to the sword. Yet I will spare a remnant, so th[at] in My anger and My turning away from them, they will not be des[royed]. (11) And the Angels of Mas[t]emoth will rule over them and ... they will turn aside and (12) do . ... what I consider Evil, walking in the stub[borness of their hearts ...]

Fragment 2

Column 1

(2) [My] house [and My altar and] the Hol[y] Temple ... (3) thus it will be done ...[flor these things shall come upon them ... and (4) the rule of Belial will [be] upon them, and they will be delivered to the sword for a week of year[s... From the] beginning of that jubilee they will (5) break all My Laws and all my Commandments that I commanded th[em, though I send them] my servants the Prophets. (6) And they wi[ll be]gin to quarrel with one another. Seventy years from the day when they broke the [Law and the] Covenant, I will give them (7)[into the power of the An]gels of Mastemoth, who will rule them, and they (i.e., the people) will neither know nor understand that I am angry at them because of their rebellion, (8)[because they aban]doned Me and did what was evil in My eyes, and because they chose what displeases Me, overpowering others for the sake of Riches and profiteering (9)... They will rob their neigh[b]ors and oppress one another and defile My Temple (10) ... and] My festivals... through [their] children they will pollu[te] their seed. Their priests will commit violence...

Column 2

(4) from it ...(5) and with a word ...(6) we ... (7) they will know, and I will send ... (8) and with compassion to as[k...] (9) in the midst of the land [and] on ...(10) their possession and they will sacrifice in it ... (11) they will pollute it and the alta[r...]

Pseudo-Jeremiah (4Q385)

This text, attributed to the prophet Jeremiah, contains many interesting characteristics. In the first place, one should note the seemingly interchangeable references to the Lord and *God*. The emphasis on 'keeping the Covenant' encountered above is continued - in this case even in captivity. The style shifts in the second and third fragments if in fact all the fragments are part of the same document - to the first person, where it would appear to become more of a pseudo-

Ezekiel than a pseudo-Jeremiah composition, though one could even opine this for the curious historical information presented in Fragment 1.

Text

Fragment 1

(1) . .. Jeremiah the Prophet before the Lord (2)[... wh]o were taken captive from the land of Jerusalem, and they went (3)... Nabuzaradan the captain of the guard (4)... and he too[k th]e vessels of the House of *God* and the priests (5) [...and] the children of Israel and brought them to Babylon. And Jeremiah the Prophet went (6)...the river, and he commanded them concerning what they were to do in the land of their captivity (7)...to the voice of Jeremiah, concerning the things that *God* commanded him (8)... and they will keep the Covenant of the *God* of their fathers in the la[nd of] (9)[their captivity ...]that they did, they and their kings and their priests (10)... *God* ...

Fragment 2

(1) from following Me, nor did his heart become too proud to serve M[e...] (2) and his days were completed, and Solomon [his son] sat [on his (David's) throne...] (3) and I gave the soul of his enemies in exchange for the soul ... (4) and hook the witnesses of Evi[l...]

Fragment 3

(1)...(2)... the Lord, and all the people arose and sa[id ...] (3) the Lord of Hosts, and I also ... her people ...(4) And the Lord said to me, 'Son of [man...] *God*... (5) they shall sleep unti[l...] (6) and from the land ... (7) he was rendered guilty...

Second Ezekiel (4Q385-389)

Beginning in Fragment 1 with a more or less familiar vision of Ezekiel's Chariot, in succeeding fragments it moves into more apocalyptic and eschatological themes. In Lines Off. of Fragment 3 Column 1, the wellknown 'bones' passage from Ezekiel is evoked with an obviously even greater emphasis on the idea of resurrection encountered in several texts above and associated with these passages from Ezekiel in the popular mind.

For instance, the 'bones' passage from Ezekiel, also used the tell-tale words 'stand up' we have encountered above, and was found buried under the synagogue floor at Masada, probably not without reason.

The text now moves on to more historical allusions, including a mysterious one to 'a son of Belial'. The language it uses in Fragment 3, Column 3 is typically that

The Dead Sea Jesus

found in other Qumran texts. There is the reference to 'Downtrodden' (Dal) and 'cup' imagery denoting, as we have seen above, that divine vengeance so important to the Habakkuk Pesher and recapitulated also with reference to Babylon in Rev. 14:8- 11.

This 'cup' imagery in the Habakkuk Pesher is extremely important, because it has been mistaken by many commentators as denoting drunkenness - the drunkenness of the Wicked Priest. But this is totally inaccurate. It actually denotes, as here, that divine vengeance being visited on the Wicked Priest for his destruction of the Righteous Teacher and his colleagues.

These references to an Era of Wickedness dominated by 'the Angels of Mastemoth' and to both a 'blasphemous king' and a 'son of Belial' increases these connections and the portentous quality of the text. It has been claimed that allusions from Second Ezekiel reappear in the Epistle of Barnabas, a second-century work brimming with the same kinds of references as this collection. The Epistle of Barnabas is so full of allusions like 'the Way of Light', 'the Way of Darkness', 'the Way of Holiness', 'the Way of death', 'keeping the Law', 'Righteousness', 'the Last judgement', 'uncircumcized heart' and 'the Dark Lord' (paralleling both Belial and mastema above) that is would be difficult not to find parallels.

Text

Fragment 1

(1) and my people shall be ... (2) with contented heart and with wil[ling soul...](3) and conceal yourself for a little while ...(4) and cleaving ...(5) the vision that Ezek[iel] saw ... (6) a radiance of a chariot, and four living creatures; a living creature [...and they would not turn] backwards [while walking;] (7) each living creature was walking upon two (legs); [its] two le[gs ...](8) was spiritual and their faces were joined to the oth[er. As for the shape of the] (9) fa[ces: one (was that) of a lion, and on]e of an eagle, and one of a calf, and one of a man. Each [one had the hand of] (10) a man joined from the backs of the living creatures and attached to [their wing.] And the whe[els...] (11) wheel joined to wheel as they went, and from the two sides of the whe[els were streams of fire] (12) and in the midst of the coals were living creatures, like coals of fire, [torches, as it were, in the midst of] G& the whe[e]ls and the living creatures. [Over their heads] was [a firmament that looked like] (14) the terrifying]ice. [And from above the firmament] came a sound ...

Fragment 2

(1)... in place of my grief... [And my heart] (2) is in confusion, together with my soul.
But the days will hasten on fast, until [all humankind] will say, (3) 'Are not the days hurrying on in order that the children of Israel may inherit [their land?'] (4) And the Lord said to me: 'I will not re[fu]se you, Ezekiel. Behold, I will me[as]ure

[the time and shorten] (5) the days and the year[s...] (6) a little. As you said to...' (7) For the mouth of the Lord has spoken these things. Fragment 3 Column 1 (1)[And I said: 'Lord, I have seen many men from Israel who] loved your Name (2)[and walked in the Ways of Righteousness. And these things, when will they happen, and] how will their Piety be rewarded?' (3) [the Lord said to me, 'I shall show t]he children of Israel, so that they will know (4) [that I am the Lord.' Then he said, 'Son of Man, pro]phesy over the bones (5)[and say: 'Draw together, bone to its bone and] joint to its joint.' And it was (6)[so. Then he said a second time, 'Prophesy, and let flesh cover the]m, and let them be covered with skin (7) [from above...]and let sinews come upon them.' (8) [And it was so. Then he said again, 'Prophesy t]o the four winds (9) [of Heaven, and let the winds blow on them, and th]ey will stand up- a great peo[ple], men...'

Fragment 3

Column 2

(1) and they will know that I am the Lord.' And he said to me, 'Consider carefully, (2) Son of Man, the land of Israel.' And I said, 'I see, Lord; it is desolate. (3) When will you gather them together?' And the Lord sai[d], 'A son of Belial will plan to oppress my people, (4) but I will not allow him to do so. His rule shall not come to pass, but he will cause a multitude to be defiled (and) there will be no seed left. (5) The mulberry bush will not produce wine, nor the bee honey ...(6) I will slay the Wicked in Memphis, and leading My sons out of Memphis, I will turn upon the re[s]t. (7) Just as they will say, 'Pe[a]ce and quiet is (ours),' so they will say 'The land r[es]ts quietly.' (8) Just as it was in the days of... ancient..., so... (9) [in the fo]ur corners of heaven... (10) [like a] consuming [fi]re ...

Column 3

(1).. . nor shall he have mercy on the Downtrodden, and he shall go to Babylon. Now, Babylon is like a cup in the Lord's hand; like refu[se] (2) he will hurl it... (3) in Babylon, and it will be ... (4) the dwelling of your fields . ..(5) their land will lie desolate ...

Fragments 4-6

(2)... and sovereignty will devolve upon the Genti[les] for [m]any[years,] while the children of Israel [1...](3)a heavy yoke in the lands of their captivity, and they will have no Deliverer, (4) because ...they have rejected My Laws, and their soul has scorned My teaching. Therefore I have hidden (5) My face from [them, until] they fill up the measure of their sins. This will be the sign for them, when they fill up the measure of (6) their sin ...I have abandoned the land because they have hardened their hearts against Me, and they do not kno[w] (7) tha [t...they have] done Evil again and again . .. (8) [and they broke My Covenant that I had made] with Ab[raham, I]saac and (9) [Jacob. In] those [days] a blasphemous king will arise among the Gentiles, and do evil things ... (10) Israel from (being) a people. In his days I will break the Kingdom (11) of Egypt... both Egypt and Israel will I break, (and) give (them) over to the sword. (12)... [hi]gh places of the 1[and ...] I have removed (its) inhabitants and abandoned the land into (13) the hands of the Angels of Mastemoth (Satan/Belial). I have hidden [My face from Is]rael. This will be their (14) sign: in the day when they leave the land ...(15) the priests of Jerusalem to serve other *God*s ... (16) three [kings] who will rul[e ...

Pseudo-Daniel (4Q243-245)

The pseudo-Daniel portions that follow describe one or more occasions on which Daniel stood before King Belshazzar (cf. Dan. 5). Like Pseudo-Jeremiah, Second Ezekiel, and the Damascus Document, they furnish a tantalizingly mysterious and often apocalyptic view of history. This text refers to: 1. the flood and the tower of Babel; 2. the exodus from Egypt; 3. the exile to Babylon; 4. the first four kingdoms (see also the Vision of the Four Kingdoms below); 5. seemingly the Hellenistic era; and 6. probably the Roman era of the 'last days' or 'end of time'.

Text

(1)... Daniel befo[re...] (2) Belshazzar ...(4) after the flood ... (5) Noah from Lubar [the mountain ..] (6) a city ... (7) the tower; [its] heig[ht ...] (9)[up]on the tower and ...(10) to visit the sons of ...(12) [fo]ur hundred [years ...] (13) ... all of them, and they will go out from (14) Egypt by the hand of ...and their crossing [will be](at)[the] River Jordan ... (15)and their sons... (17) the children of Israel preferred their presence (i.e., that of the false *God*s) to the [presence of *God*.] (18) [They were sacrific]ing their sons to the Demons of Falsehood, and *God* was angry at them and de[tided] to give (19) them into the power of Nebu[chadnezzar the king of Ba]bylon, and to lay waste to their land before them by the hands of ...(2 0) members of the exilic community... (21) and He scattered them... (22) an oppression of seve[nt]y years ...(23) this great [kingdom], and He will save the[m...] (24) powerful, a Kingdom of the Peop[les...] (25) this is the fi[rst] Kingdom ... (26) [he will] rule (some number of) years... (27) Balakros ... (29) [y]ears ... (30) ... rhosthe son of ... (31) [and ... rh]os [will rule] thirty-five years... (32) to say... (33) [Ev]i[l] has led astray... (34) [in] this [time] the called ones will be gathered .. . (35)[the Kings of] the Peoples, and from (that) day on there shall be ... (36)[Holy] Ones and the Kings of the Peoples ... (37) they shall be doi[ng] until [this] day .. .(39) and what (40). .. Daniel (41) ... a book that he gave (42)... Kohath (43) ... Uzzi[ah...] (44) A[b]iathar... (46) Jehoniah (47) ... Simeon (48) ...David, Solomon (49) ... Ahazi[ah, ...] (51) to bring Evil to an end (52)... these will wander astray in blindness (53)... [th]ese (people.) Then there shall arise (54)... [H]oly O[n]es shall return (5 5)... Evil.

The Son of God (4Q246)

This is another Messianic pseudo-Daniel fragment in Aramaic, relating to the literature centering about that figure. It is full of the language and heightened imagery of these apocalyptic visionary recitals. In fact, it takes its cue from a reference in the Biblical Daniel to the 'Kingdom' that 'the *God* of Heaven will set up ... which shall never be destroyed', nor be conquered or absorbed, but rather 'last forever' (Dan. 2:44).

God, War Scroll, etc.., whether taken figuratively or otherwise, is extremely warlike. This is in line with the general uncompromising, militant and nationalist ethos of the Qumran corpus; the Messianic figure was to be a triumphant, quasi-nationalist king figure. One should also note that the peace envisaged in this text will only come after the cataclysmic Messianic war. As in the War Scroll, *God* will assist in this enterprise with His Heavenly Host. For the War Scroll, this is the point of the extreme purity regulations and being in camps in the wilderness, which is put in vii-5-6 as follows: 'because the Angels of Holiness are with their hosts,' i.e. 'the war volunteers, the Perfect in Spirit and body ready for the Day of Vengeance'. We shall learn more about the extreme purity regulations required in the 'camps' in the last column of the Damascus Document in Chapter 6 below.

Text

Column 1

(1) [the king. And when the Spirit] came to rest upo[n] him, he fell before the throne. (2) [Then Daniel arose and said,] 'O Wing, why are you angry; why do you [grind] your teeth? (3)[The G]rear [*God*] has revealed to you [that which is to come.] It shall indeed all come to pass, unto eternity. (4) [There will be violence and gr]eat [Evils.] Oppression will be upon the earth. (5) [Peoples will make war,] and battles shall multiply among the nations, (6)[until the King of the people of *God* arises. He will become] the King of Syria and [E]gypt. (7) [All the peoples will serve him,] and he shall become [gre]at upon the earth. (8)[... All w]ill make [peace,] and all will serve (9) [him.] He will be called [son of the Gr]eat [*God*;] by His Name he shall be designated.

Column 2

(1) He will be called the son of *God*; they will call him son of the Most High. Like the shooting stars (2) that you saw, thus will be their Kingdom. They will rule for a given period of year[s] upon (3) the earth, and crush everyone. People will crush people, and nation (will crush) nation, (4) until the people of *God* arises and causes everyone to rest from the sword. (5) His Kingdom will be an Eternal Kingdom, and he will be Righteous in all his Ways. He [will jud]ge (6) the earth in Righteousness, and everyone will make peace. The sword shall cease from the earth, (7) and every nation will bow down to him. As for the Great *God*, with His help (8) he will make

war, and He will give all the peoples into his power; all of them (9) He will throw down before him. His rule will be an Eternal rule, and all the boundaries...

The Vision of the Four Kingdoms (4Q547)

This is another tantalizing apocalypse in Aramaic relating to the Daniel cycle of literature, as well as to a certain extent Enoch. In it, the king (possibly either Belshazzar or Nebuchadnezzar) sees a vision of four trees, each represented by an Angel. As each tree represents a kingdom, some relationship with the Dan. 7-8 vision of the four kingdoms is evident.

When the text assigns Angels to trees, and thereby to the kingdoms, it is developing an already ancient idea prominent in Daniel. In Dan. 10:13, the seer encounters an Angel, presumably the Heavenly interpreter of visions, Gabriel. This Angel is also of fundamental importance to the heir to many of these traditions, Islam. He tells Daniel that he would have come earlier, but 'the prince of the Kingdom of Persia opposed' him for 2 1 days.

Text

Fragment 1

Column 1

(5) ... the [Li]ght of the Angels who were (6) ... he said to them, 'It will all happen ...'(7) high. It is he who ... (8) and he said to me, 'O King, because...' (9) as everything was done, they would arise (10)... he said, 'They shall be.' And he explicated to them clearly (11) ... their lords. One of them (12) [... Then the Angel upon w]hom Fragment 1

Column 2

(1) (rested) the brilliant Light arose, and the four tree[s after him.] (2) A tree arose, and (the others) moved away from it. He (the Angel) sai[d to me ...'What](3) kind of tree is it?' I replied, 'If only I could see and underst[an]d it.' [Then I saw] (4) a balsam tree... (5) I asked it, 'What is your name? 'It replied, 'Babylon.' [Then I said to it,] (6) '[Yo]u are he who shall rule over Persia.' Then [I saw another tr]ee (7) [that was be]low where we were standing, and it swore... and claimed (8) to be different. (superior to the previous tree?)... So I asked him, 'What is [your] name?' [He replied ...] (9) I said to him, 'You are he wh[o shall rule over...'] [By] (10) my power and by the region... he swore [... And I saw] (1 1) [the] third tree, [and] I said to [him, 'What is your name?' He replied,] (12)'Your vision ...'

Fragment 2

(1) ... and I said to him, 'This is it. Who is the ruler of...'

Fragment 3

(9)... the lord of ... (10) the Most High *God* ... (11) which is upon them ... (12) [the Lor]d of all, he who appoints judges ...

Fragment 4

(9) ... they shall seize (10) ... the vision (11)... you have spoken (12) [...the kin]g who shall escape ...

Testaments and Admonitions

Fernando Klein

Aramaic Testament of Levi (4Q213-214)

Even though it is possible to harbour reservations about whether all this material of varying emphases can, in fact, ever be made to correspond to a single whole, it is important to note themes and imagery even in well-known texts such as this one, which at once move across the entire spectrum of Qumran literature and are completely harmonious with the Qumran perspective.

Here we have the typical emphases on 'Righteousness', 'Truth', 'Judgement', 'Knowledge' and 'Wisdom' as opposed, for instance, to 'Evil' and 'fornication'. The 'Righteousness' and 'fornication' themes at Qumran are, as we have seen, particularly strong. Though it is perhaps possible to identify themes such as these in literature of the Second Temple period generally, their emphasis in documents like this one is particularly telling. It is also interesting to see the descendants of Levi or the priests denoted in 1.2.2 as 'a Righteous seed'.

Text

Manuscript A

Fragment 1

Column 1

(8)... this (9)... I (10)... I [washed myself] and all (11) [...then] I raised [my eyes and my countenance] to Heaven (12)... my toes and my fingers (13) [...1 prayed and] said, 'My Lord, You (14) know... You alone know (15) [...all] the paths of Truth. Put away (16) [from me... Evil and fornication. Turn away (17) [...Wi]sdom and Knowledge and strength (18) [give to me... [to] find Your Mercy before You (19)... that which is pleasant and good before You (20)... let [no]t any satan (here, possibly 'enemy') have power over me (21) [...up]on me, Lord, and bring me forward to be Your...

Column 2

(8) [Your] a[yes...] (9) Lord, [You] bl[essed...] (10) a Righ[teous] seed. [... Hear, please,] (11) the prayer of [Your] ser[vant Levi... to do] (12) a True judgement for al[l time... Do not remove] (13) the son of Your servant from [Your] pre[sence...] (14) Then I continued on... (15) to my father Jacob and whe[n...] (16) from Abel Mayin. Then (17) I lay down and I remained a[t Abel Mayin...] (18) Then I was shown visions... (19) in the vision of visions, and I saw Hea[ven opened and I saw the mountain] (20) beneath me, as high as to reach to Heav[en, and I was on it. Then were opened] (21) to me the gates of Heaven, and an Angel [spoke to me, 'Levi, enter...']

Fragment 3

(1)... your priesthood above all flesh (2) [...And] I awoke from my sleep. Then (3) [I said, 'This is a vision, and thus I am amazed that I should have any vision at all.'] I [hid] this also in my heart; to no person did I (4) [reveal it. And we went to my (grand)father Isaac, and he also blessed me thus. Then, wh]en Jacob my father was tithing (5) [everything that he owned according to a vow made to *God*, for the first time I was at the head of the priest]s, and to me of (all) his sons he gave...

Fragment 4

Column 1

(1) [And in the hundred and eighteenth year of my life, the year] in which died (2) [my brother Joseph, I called my children and their children and began to instruct] them (3) [concerning all that was in my heart. I said to my children, 'Listen] to the word (4) [of Levi your father, and pay heed to the instructions of *God*'s friend.] I [instruct] you, (5) [my sons, and I reveal the Truth to you, my beloved. The essence] of each of your works (6) [must be Truth. May Righteousness always] re[main] with you, and Truth. (7) [Then you shall have a blessed and good [harvest.] He who sows Good reaps Good, (8) [while he who sows Evil,] his [sowing turns against him.] And now, my sons, [teach] Torah, its interpretation and Wisdom (9) [to your sons, and Wisdom shall be with you] as an eternal honor. He who teaches Wisdom (10) [will find] honor [therein. He who despises Wisdom] will be given [to con]tempt and disdain. Observe, my sons, (11) [Joseph my brother, who taught Torah and interpretation and Wisdom. (He received) honor and became a great man, both to kings (12)... Do not exchange Wisdom for a teacher (13)... a foreigner. He who teaches Wisdom, all (14) [the days of his life shall be long, and his reputat]ion [shall grow great.] In every land and country to which he goes (15) [he has a brother, and is not like a stranger] in it, nor like a foreigner in it, neither (16) [is he like an unfamiliar person there. For all give him honor there, because everyone wants (17) [to learn from his Wisdom.] His [friends] are numerous, and many seek his welfare. (18) [They seat him on the seat of honor, in order] to hear his words of Wisdom. (19) [Wisdom is a great richness of honor for] those who know it, and it is a treasure (20) [for everyone who possesses it. If] migh[t]y [kings come] with a [great] (and) powerful army...

Column 2

(1) (they will not find) its hidden places, and they shall not enter its gates, nor... (2) will they be able to conquer its walls. And not... (3) they will see its fodder, its treasure that does [not]... (4) for which there is [n]o price... [Any man] (5) who seeks Wisdo[m, Wisdom [will find] him... [it will not] (6) be hidden from him... (7) and not la[c]king... all who see[k] (8) in Truth... Torah and interpretation, (9) Wi[sd]om that [they] teach... they are two (10)... great. You will give (11) [...h]onor... (12) Also in books (13) [...you will b]e leaders and judges (14)... and

servants (15)... also priests and kings (16)... your Kingdom (17) will l[ast forever... there will be no end (18)... [the Kingdom will not] pass away from you until (19)... in great honor.

Fragment 5

(1) [...y]ou will be darkened... (2) did not [E]noch complain... (3) Noah, and upon whom does the guilt fall?... (4) is it not upon me and upon you, my sons? Now, you know (5)... the Ways of Righteousness you will aba[n]don, [and] all the ways (6)... you will renounce, and you will walk in Darkn[ess] (7)... great oppression will come upon you, [and] you will be gi[ve]n (8)... you will become fools...

Manuscript B

Fragment 1

(1) to offer up on the altar (anything fitting), wash your hands and feet once again. And offer] split [woo]d. (2) [Examine it first for an[y worms, and t]he[n offer] it up, [for] thus I saw Abraham (3) [my father taking care regarding] anything that might restrain (him from offering the wood). Any of [twel]ve woods [that are fitting] he to[ld] me [to] offer up on the altar, (4) [whose smoke] rises up with a pleasant odor. These are th[eir] na[mes:] ce[dar, juniper, almond, (5) [fir, pine, ash, cypress, fig, oleaster, laurel, myrtle and (6) as[ph]althos. These are th[ose he said] are fitt[ing to offer up] under the burnt offering (7) on the altar. And [when you begin to offer up one of these [wood]s upon [the] altar and the [fire begins] (8) to burn them, you are to sprinkle the blood] on the sides of the altar. Again, [wash] (9) [your hands and feet of the blood, then begin to offer up] sa[lt]ed portion[s.] As for [its[he[ad,]...

A Firm Foundation (AARON A - 4Q541)

The relationship of this text to the Aramaic Testament of Levi above should be clear. In fact, there is no real reason to consider it distinct from it, but rather simply another version or portion of it. The working title, 4QAaron, must be seen as a convention, nothing more, though it does reflect the priestly character of some of the material, particularly the references in Column 2.4 to 'burnt offerings' and being 'a Foundation of peace' and similar allusions, including more 'Foundation' imagery in Columns 4 and 6.

The reconstruction involving 'crucifixion' in Column 6 is also interesting, if it is finally to be entertained. Here, combined with the 'firm Foundation' imagery, it moves into a splendid evocation of eternal life in terms of reference to the 'Light' imagery so widespread at Qumran and, of course, the beginning of the Gospel of John. This imagery has much in common with that of the Testamental bequests and recitations associated with Kohath and Amram.

The Dead Sea Jesus

Text

Column 1

(Fragment 6) (1)... dee[p things]... (2) who doesn't understand. And he wro[te]... (3) and he stilled the great sea... (4) Then the books of Wis[dom] will be open[ed]... (5) his word... Column 1 (Fragment 2) (1)... wo[rds]... and according to the will of (2)... to me. Once more he wrote (3)... I [sp]oke concerning it in parables (4)... was near to me. Therefore, was far from me (5)... The visi[on] will be [profou]nd... .the fruit ...

Column 2

(2) [from] *God*... (3) You shall receive the affli[cted ones]... (4) [You] shall bless [their] burnt offerings [and You shall establish for] them a Foundation of Your peace... (5) your Spirit, and you will rejoice [in your *God*. Now] I [am proclaiming to you parable[s]... rejoice. (6) Behold, a wise man [will understand that I am seeing] and comprehending deep Mysteries, thus I am spec[king...] parable[s]. (7) The Greek (?) [will not understand. But the Knowledge of Wisdom will come to you, for you have received... [you] will acquire... (8) Pursue her (Wisdom) and seek [her and gain possession of] her to swallow (her) down. Behold, you will gla[dden] many... many (will have) a place...

Column 4

(1)... his Wisdom [will be great.] He will make atonement for all the children of his generation. He will be sent to all the sons of (2) his [generation]. His word shall be as the word of Heaven and his teaching shall be according to the will of *God*. His eternal sun shall burn brilliantly. (3) The fire shall be kindled in all the corners of the earth. Upon the Darkness it will shine. Then the Darkness will pass away (4) [from] the earth and the deep Darkness from the dry land. They will speak many words against him. There will be many (5) [lie]s. They will invent stories about him. They will say shameful things about him. He will overthrow his evil generation (6) and there will be [great wrath]. When he arises there will be Lying and violence, and the people will wander astray [in] his days and be confounded.

Column 5 (Fragment 5)

(1)... and those who are grieved concerning... (2) your ju[dgment] but you will not be gui[lty]... (3) the scourging of those who afflict you... (4) your complaint (?) will not fail and all... (5) your heart be[fore]...

Column 5 (Fragment 1)

(1) [Behold] I saw one... (2) I saw seven rams... (3) Some of his sons shall walk... (4) They shall be gathered to the Heav[enly Beings]...

Column 6

(1) *God* [will set] right error[s]... [He will judge] revealed sins... (2) Investigate and seek and know how Jonah wept. Thus, you shall not destroy the weak by wasting away or by [crucif]ixion... . (3) Let not the nail touch him. Then you shall raise up for your father a name of rejoicing and for all of your brothers a [firm] Foundation. (4)... You shall see and you shall rejoice in the Eternal Light and you will not be one who is hated (of *God*).

Testament of Kohath (4Q542)

This text belongs to the genre of pseudepigrapha like the Testament of Levi and A Firm Foundation (Aaron A) above. All of these texts, including the one attributed to Amram below, are associated with one or other of the principal characters in the priestly succession; in fact, we may be witnessing a tradition here not dissimilar to that of the Rabbinical one (e.g. the Pirke Abbot), where one after another of the important Rabbinical figures in the line of succession had important sayings attributed to him. Instead of a succession of rabbis, however, this one would have consisted of priestly forebears.

Text

Fragment 1

Column 1

(1)... and *God* of *God*s for all Eternity. And He will shine as a Light upon you and He will make known to you His great Name (2) and you will know Him, that He is the Eternal *God* and Lord of all creation, and sovereign (3) over all things, governing them according to His will. And He will give you joy and your sons rejoicing for generations of (4) the Truth, forever. And now, my sons, be watchful of your inheritance that has been bequeathed to you, (5) which your fathers gave you. Do not give your inheritance to foreigners, nor your heritage to (6) violent men, lest you be regarded as humiliated in their eyes, and foolish, and they trample upon you, for (7) they will come to dwell among you and become your masters. Therefore, hold fast to the word of Jacob (8) your father, and be strong in the judgements of Abraham, and in the Righteousness of Levi and myself. Be Holy and pure (9) from all vacat entirely, and hold to the Truth, walking uprightly without Deceitfulness, (10) but rather with a pure heart and in a True and Good Spirit. Thus you will grant to me a good name among you, together with joy (11) for Levi and happiness for J[a]cob, rejoicing for Isaac and blessing for Abraham, inasmuch as you guarded (12) and walked (in) the inheri[tance. My solos, your fathers bequeathed to you Truth, Righteousness, Uprightness, (13) integrity, pur[ity,

Holiness and the priesthood. In accordance with what you have been commanded, and according to all that

Column 2

(1) I have taught you in Truth, from now and for al[l Eternity... (2) the words of the Truthful saying. There will come upon yo[u... (3) eternal blessings shall rest upon you. And there shall b[e... (4) enduring to all the generations of Eternity. And no longer shall you... (5) from your Foundation, and you shall endure to pronounce judgements... (6) to reveal the sin of all the eternal sinners... (7) and the Wicked and in the depths of the sea and in all the hollows of the earth... (8) in [gener]ations of Truth, while all the sons of Evi[l] will pass away... (9) And now, you, Amram, my son, I appoi[nt... (10) and [to] your [son]s and to their sons I appoint... (11) and they bequeathed it to Levi my father, and Levi my father [bequeathed it] to me... (12) and all my books as a testimony, that you take heed of them... (13) Great merit [will come] to you from them as they accompany you in your affairs.

Fragment 2

Column 1

(6)... his sons... (7) among mankind and among liv[ing things]... (12) a Light. Instead... (13) and I . ..

Fragment 3

Column 1

(in the margin) his root.

Column 2

(11)... the stones will call... (12) them from all fornication very much. Whoever... (13) exceedingly, for he has no...

Testament of Amram (4Q543, 545-548)

The Testament of Amram, if indeed we can call it this - Amram per se is mentioned only in Manuscript C - is one of the most splendid apocalyptic and visionary works in the corpus. In it, many of the themes we have encountered in the works discussed above come together in a fairly rationalized eschatological whole.

These include the usual 'Light', 'Darkness', 'Belial', 'Righteousness', 'Truth', 'Lying' and 'Watcher' vocabulary, including the very nice allusions to 'sons of Righteousness' - which we have already identified as a variant of the 'sons of

Zadok' terminology - 'sons of Light', 'sons of Darkness' and 'sons of Truth', again widely disseminated through the whole corpus.

Text

Manuscript C

Column 1

(1) A copy [of the book of the words of the vis]ion of Amram the son of Kohath, the son of Levi: all (2) that [he revealed to his sons and that he commanded t]hem on the day [of his death, in] the year (3) one hundred and thirty-six, whi[ch] was the year of his death, [in the o]ne hundred (4) and fifty-second year of the exil[e of Israel to Egypt... upon [him, and he sent] (5) to call Uzziel his youngest brother, and [gave] to him [in marr]iage [Miri]am [his] daughter. [For he said] (6) 'You (Miriam) are thirty years old.' Then he gave a wedding feast seven [day]s long, (7) and ate and drank and rejoiced at the feast. Then, when (8) the Mays of the wedding feast were completed, he called for Aaron his son. Now, [h]e (Aaron) was a man of... years of age (9) [...and he said] to him, 'Call... and Malachijah... from the house of (10)... above. He called him... Column 2 (11) in this land, and I went up to... (12) to bury our fathers. And I went up... (13) to [a]rise, to bind and pile sheaths and to build... (14) gre[a]t from the sons of my uncle, all togeth[er... and from] (15) our excee[ding great labors, [until in Egy]pt there died... (16) the rumor of war and unrest returned... to the land of E[gypt...] (17) to meet and [they had] not bui[lt gr]aves for their fa[th]ers. Then [my father Kohath] released me [to go,] (18) to build and to get [all their needs] for the[m from the land of Canaan. [And while] (19) we [were] building, wa[r broke out between] the Philistines and Egypt, and... was winn[ing...]

Manuscript E

Fragment 1

(1)... that Levi his son sacrificed to... (2) I said to you at the alt[ar] of stop[e...] (3) [con]cerning sacrifice[s...]

Fragment 2

(2) I rescued... (3) he built... (4) at Mount Sinai... (5) a great blessing at the bronz[e] altar... (6) from among all the people on earth his [son] will be exalted as a priest. The[n...] (7) and his sons after him, for all the generations of eternity in Tru[th...] (8) and I awoke from the sleep of my eyes, and [I] wrote down the vision... (9) I went out from the land of Canaan, and it happened to me just as he said... (10) exalting, and afterwar[d], in the twen[tieth] year, [I returned to the land of Canaan] (11)... you were...

The Dead Sea Jesus

Manuscript B

Fragment 1

(9) [...I saw Watchers] (10) in my vision, the dream-vision. Two (men) were fighting over me, saying... (11) and holding a great contest over me. I asked them, 'Who are you, that you are thus empo[wered over me?' They answered me, 'We] (12) [have been em]powered and rule over all mankind.' They said to me, 'Which of us do yo[u choose to rule (you)?' I raised my eyes and looked.] (13) [One] of them was terr[i]fying in his appearance, [like a serpent, [his] cl[oa]k many-colored yet very dark... (14) [And I looked again], and... in his appearance, his visage like a viper, and [wearing...] (15) [exceedingly, and all his eyes...]

Fragment 2

(1) [...em]powered over you... (2) [I replied to him,] 'This [Watcher,] who is he?' He answered me, 'This Wa[tcher...] (3) [and his three names are Belial and Prince of Darkness] and King of Evil.' I said, 'My lord, what dom[inion...?'] (4) ['and his every way is darkened, his every work da[rk]ened. In Darkness he... (5) [Yo]u saw, and he is empowered over all Darkness, while I [am empowered over all light.] (6) [...from] the highest regions to the lowest I rule over all Light, and over al[l that is of *God*. I rule over (every) man]

Fragment 3

(1) [of His grace and peace. Over all the sons of Lig]ht have] I been empowered.' I asked him, [What are your names... ?'] (2) He [s]aid to me, '[My] three names are [Michael and Prince of Light and King of Righteousness.']

Manuscript ?

Column 1

(1)... tribes... (2) to them and all [his] ways [are True...] (3) [and he will heal] them of all their ills... (4) them from death and from de[struction...] (5) [o]ver you, blessed sons... (6) all the generations of Israel for[ever...] (7) angry at me, for the sons of Ri[ghteousness...] (8) between the sons of Lying and the sons of Tr[uth...] (9) I [will make known to you;] certainly I will inform y[ou that all the sons of Light] (10) will be made Light, [whereas all the sons] of Darkness will be made Dark. [The sons of Light...] (11) and in all their Knowledge [they will] be, and the sons of Darkness will be dest[ro]yed... (12) For all foolishness and Evi[l will be darkened, while all [pea]ce and Truth will be made Ligh[t. All the sons of Light] (13) [are destined for Light and [eternal j]oy (and) [re]joic[ing.] All the sons of Dark[ness] are destined for [Darkness and death] (14) and destruction... Lightness for the people. And I shall reveal [to you...] (15) from Darkness, for all... (16) the sons of [Darkness...] and all the sons of Light...

189

Testament of Naphtali (4Q215)

The Testament of Naphtali has long been known in its Greek form, which like the Testament of Levi is part of the apocryphal Testament of the Twelve Patriarchs. The original Hebrew or Aramaic version (it was not known which it might be) had, it was long thought, perished in antiquity. Then, in 1894 M. Gaster drew the attention of the scholarly world to medieval manuscripts that contained two slightly different Hebrew versions of the work. The relationship between the Greek and the Hebrew has been a matter of some debate ever since.

The surviving portions of the Qumran version of the work, presented here, will doubtless fuel the debate still further. Column 2 preserves the Hebrew form of the Greek Testament of Naphtali 1:9, 11- 12, along with previously unknown details. Column 4 does not parallel any portion of the Greek, and has an eschatological thrust not found in the Greek Testament of Naphtali.

Text

Column 2 (or later)

(1) with the paternal aunts of Bilhah, my m[other, and her sister, Deborah, who nursed Rebeccah... (2) And he (the father) went into captivity, but Laban sent and rescued him. Then he (Laban) gave him (Rotheos) Hannah, one of his maidservant[s. And she conceived and bore] (3) her first [daughter], Zilpah. And he gave her the name Zilpah after the name of the city to wh[ich] he had been taken captive. (4) Then she conceived and bore Bilhah, my mother. Hannah named her Bilhah because, when she was born, (5) she hurried to suckle. So she said, 'My, how my daughter is in a hurry.' So from then on she was called Bilhah. (6) vacat (7) And when Jacob my father, fleeing from Esau his brother, came to Laban, and because [Rotheos], (8) the father of my mother Bilhah, di[ed], Laban took charge of Hannah my grandmother and her two daughters. He gave one daughter (9) to Leah and one to Rachel. When it came about that Rachel was not bearing sons... (10) Jaco]b, my father. And Bilhah, my mother, was given to him; so she bore Dan [my] brother [...(11) tw]o daughters...

Column 4 (or later)

(1) [...in] the Pit (2) and great distress and devilish trials. And some among them shall be purified to become the Elect of Righteousness... (3) for the sake of His Pious [On]es. For the era of Evil has been completed, and all sinfulness will pas[s away]; (4) the time of Rightousness has come, and the earth will be full of Knowledge and praise of *God*. F[or] (5) the era of peace has come, and the laws of Truth and the testimony of Rightousness, to teach [all mankind] (6) the Ways of *God* and the mightiness of His works; [they shall be instructed until all Eternity. All cr[eation] (7) will bless Him, and every man will bow down before Him in worship, and their he[arts will be] as one. For He [prepared] (8) their actions before ever

they were created, and [measured out] the service of Righteousness as their portion (9) in their generations. For the rule of Goodness (Righteousness) has come, and He has raised up the Throne of the [Messiah.] (10) And Wisdom will increase greatly. Insight and understanding will be confirmed by the works of [His] Holiness...

Admonitions to the Sons of Dawn (4Q298)

On the surface, there is nothing astonishing about this text and no reason why it should be written in an unknown script, called earlier by some scholars 'cryptic'. We have been able to decipher it and provide the following chart, which equates the letters used in the text with their corresponding Hebrew forms:

Text

Column 1

(1) [The word]s of (the) Maskil (Teacher) that he spoke to all the sons of Dawn (cryptic script begins): Give e[ar to me], all men of heart (2) and those who [pus]sue Righteousness: you will und[er]stand my words and be seekers after Faith. H[ea]r my words with [all your (3) strength. Lis[ten]... [kn]owing the ways of... [ac]hieve [long] life, men of (4)... search out...

Column 2

(1) its roots reach [out]... a Glorious abode (2) in the depths be[low]... and in them. (3) Consider... dust (4)... *God* gave (5)... on all the earth (6)... he measured their setting (7)... under the name of (8)... their setting, to go about (9) ... a storehouse of Understanding (10)... and which...

Column 3

(1)... and recounting its boundaries (2)... not to be on high (3)... the Glorious abode. And now, (4) give ear... and knowing, hear. And men (5) of Understanding,... and those seeking judgement, humility, (6)... add strength. And the men of (7) Truth, pursu[e Righteousness] and love Piety; add (8) Meekness... the hidden things of the testimony, which he (9) solved... so that you understand the era (10) of Eternity, and examine the pa[s]t, so as to know... Column 5 (8)... destruction (9)... to tread... Column 6 (8) of the dawn... (9) its boundaries... (10) he placed boundaries...

The Sons Of Righteousness (Proverbs - 4Q424)

In this text, another typical 'Wisdom' text, there is the usual Qumran vocabulary of 'Judgement', 'Riches', 'Knowledge', and in Line 1.9, an additional one - 'zeal for Truth'. This is preceded and followed by an interesting additional evocation of 'deceitful' or 'cunning lips', which will also be of interest in the parallel admonitions which follow next in the Demons of Death (Beatitudes). In 1.13 there is a curious reference to 'swallowing', in this case coupled with a reference on the same line to 'the Kittim'.

Text

Fragment 1

(2)... with a fruitful man... (3) outside and chooses to build it, and he covers his wall with plaster, as well as... (4) it will fall off under the rain. Don't take legal instruction from a deceitful person and don't go, young man, with someone who is unsta[b]le. (5) For just as lead melts, so before a fire he will not stand. (6) Do not put a slackard in charge of an important task, for he will not carry out your charge. And don't send (him) to pick (7) some[thing] up, for he will not pay attention to your specifications. Do not tr[ust] the complainer (8) to get provisions for your needs. Do not trust the man of cunning lips... (9) your Judge[m]ents. He will certainly speak deviously, not being zealous for Truth... (10) with the fruit of his lips... Do not put the man with a covetous eye in charge of Rich[es...] (11) And arrange what remains to suit yourself... your dead (?)... (12) and in the time of the harvest he will be found unworthy, quick to anger... (13) Kittim, because he will surely swallow them... A man...

Fragment 2

(1) and will not do his work carefully. The man who judges before investigating is like someone who believes before ... (2) Do not give him authority over those pursuing Knowledge, for he will not understand their judgement, to justify the Righteous and condemn [the Wicked...] (3) he will also be robbed. Do not send the man of poor eyesight to look for the Upright, fo[r...] (4) Do not send the poor of hearing to seek judgement, for he will not carefully consider disputes between men. Like someone winnowing wind... [Like a...] (5) who doesn't investigate, so is someone who speaks into any ear that doesn't listen or who tries to talk to a drowsy man, who falls asleep with the spirit of... (6) Do not send the hard-hearted man to discern thoughts, because his heart's Wisdom is defective, and he will not be able to control... (7) nor will he find the discernment of his hands. The clever man will profit from Understanding. A Knowing man will bring forth Wisdom... (8) An Upright man will be pleased with judgement... A man... A soldier will be zealous for...

The Demons of Death (Beatitudes - 4Q525)

This text has been called 'the Beatitudes', comparing it to famous recitations of a parallel kind in Ecclesiasticus (Ben Sira) and the Sermon on the Mount in the Gospel of Matthew. This is perhaps a misnomer. Once again, we have a typical 'Wisdom' text here, but one also rich in the vocabulary of Qumran. Superficially, the text is fairly straightforward and commonplace. As such, it has much in common with the Sons of Dawn and Sons of Righteousness (Proverbs) materials above - at least Columns 1-3 do.

Text

Column 1

(1) [Now, hear me, all my sons, and I will speak] about that Wisdom which *God* gave me... (2) [For 13e gave me the Knowledge of Wisdom and instruc[tion] to teach [all the sons of Truth]...

Column 2

(1) [Blessed is he who walks] with a pure heart and who doesn't slander with his Tongue. Blessed are they who hold fast to her Laws and do not hold (2) to the ways of Evil. Bless[ed] are they who rejoice in her and do not overflow with the ways of folly. Blessed are they who ask for her (3) with clean hands and do not seek her with a deceitful [heart]. Blessed is the man who grasps hold of Wisdom and walks (4) in the Torah of the Most High and directs his heart to her Ways and restrains himself with her disciplines and always accepts her chastisements. (5) and doesn't cast her off in the misery of [his] affliction[s] nor forsake her in a time of trouble, nor forget her in [days of terror, (6) and in the Meekness of his soul, doesn't despis[e her], but rather always meditates on her, and when in affliction, occupies himself [with *God*'s Torah; who al]1 (7) his life [meditates] on her [and places her continually] before his eyes so he will not walk in the ways of [Evil]... (8) in unity and his heart is Perfect. *God*... (9) and W[isdom will lift up] his h[ead] and sea[t him] among kings... . (10) They [shall look upon... brothers will be fr[uitful]... (12) Now, my sons, War my voice and do] not turn aside [from the words of my mouth ....

Column 2 (Fragment 4)

(1)... to possess her with his heart... (2) with a deceitful heart. And in W[isdom]... (3) [You shall not] abandon [your inheritances to a foreign wife or your hereditary portions to foreigners, because those with Wi[sdom]... (4) They shall consider . ..(the Torah) of *God*, protect her paths and walk in [all her Ways.] (5)... her statutes, and not reject her admonishments. Those with Understanding will bring forth [words of insight... (6) (and) walk in p[eace]. The Perfect will thrust aside Evil.

They will not reject her chastisements... . [Those with Wisdom] (7) will be supported [by the strength of Wisdom]. The intelligent will recognize her Ways [and plumb] her depths... (8) The Lovers of *God* will look upon her, walking carefully within her bounds.

Column 3

(1) [No]... is like her... (2) She will not be bought with gold or [silver]... (3) nor any precious gem... (4) they resemble one another in the be[au]ty of their faces... (5) and purple flowers with... (6) crimson with every [delightful] garment... (7) and with gold and rubies...

Column 4

(2) for the atonement of sin and for weeping... (3) they shall lift up your head... (4) Perfection because of your word and Perfection... (5) for splendor and lovely in... (6) was revealed in your Ways. You shall not waiver... (7) You will be upheld at the time you falter, and you will find [Grace...] (8) The reproach of those who hate [you] shall not draw near you... (9) together, and those who hate you will be destroyed... [Shall rejoice] (10) your heart and you shall delight in [*God*]... (11) *God* [your] father has taught, and on the [backs] of your [enemies] will you tread. And... (12) Your soul shall deliver you from all Evil, and the dread of [your enemies] shall not come near you. (13) He will cause you to inherit, and fill your days with Goodness, and in abundance of peace you shall de[light]... (14) You shall inherit Glory. Even though you pass away to (your) eternal abode, [all your loved ones] shall inherit... (15) All those who know you shall walk in harmony with your teaching [and] he[ar your words]... (16) Together will they mourn and in your ways remember you, for you were... (18) And now, understand, hear me, and set your heart to [do]... (19) Bring forth the Knowledge of your inner self and in... meditate... (20) In the Meekness of Righteousness bring forth [your] words in order to give them... [Don't] (21) respond to the words of your neighbor lest he give you... (22) As you hear, answer accordingly... [Do not] (23) pour out complaints before listening to their words. And... [Do not respond] vehemently (24) before hearing their words. Afterwards respond [in the Perfection of your heart.] (25) And with patience utter (your words) and answer truthfully before officers (even 'rulers') with a To[ngue of... (26) with your lips, and guard against the stumbling block of the Tongue... (27) lest you be convicted by your lips and ensnared together with a Tong[ue of... (28) impropriety... from it and they will be perverse...

Column 5

(1)... Darkness... poison... [all] those born [on the earth]... Heaven... (2) serpents in [it, and you will] go to him, you will enter... there will be joy [on the day] the Mysteries of *God* [are revealed] for[ever]. (3)... burn. By poi[sons] will a serpent weaken his lords... [the Kingdom of *God*... [vip]ers... (4) In him they take their stand. They are accursed for[ever] and the venom of vipers... the Devil

(Mastemah)... you choose depravity... (5) and in him [in his authority] the Demons of Death take wing. At his doorway you will cry out... . [He did] Evil. He [acted wickedly. In him they exalt themselves. They walk [in his ways.] (6) [He is] your he[ad]. [From] his council there are sulph[urous] flames. And from his den are... in order to destroy those who wallow in the filth of [sin]. (7)... the reproach of disgrace, his bolted (doors) are the fasts of the pit... they increase. One calling from the wall... (8) They shall not reach the paths of life... . the walkers in the way of... in simmering anger and in pati[ence...]

Column 5 (Fragment 5)

(1)... the princes... (2) in it those who understand wander astray... (3) those who ensnare... (4) [they shed] blood. They put to de[ath the righteous]... (5) [they] act treacherously... (6) [de]ath. And the Down[trodden]...

Column 5 (Fragment 6)

(1)... might... (2) in the midst of abominations... (3) its height... (4) and in the pollution of...

Column 5 (Fragment 7)

(1)... his ser[vants] tremb[led]... (2) They filled the whole [earth with violence.]... (3) The serpent who made [every generation tremble died... . (4) [He] stationed an Angel around... (5) a Wat[cher] and G[od]...

Column 6 (Fragment 9)

(1) with/people... (2) the appointed time... (3) in Piety... (4) for every genera[tion]...

Column 6 (Fragment 10)

(1) Lest you bring forth words of [folly.]... (2) heart. Listen to me and be sti[ll before me.]... (3) I have understood, so drink from [the Well of Life]...(4) His Temple. [They walked... (5) His Temple is dwelling among... (6) forever marching... (7) or what grows of its own they shall gar[her...] (8) burned, and every weed [He uprooted...] (9) a Well of de[ep] Waters...

Fernando Klein

Sectarian Literature

War Rule (Serekh ha-Milhamah, 11Q14)

This original composition of the sect contains a vision of the "end of days" in which the sect engages in a battle with the Romans ("the kittim") and emerges victorious. Herodian script from 20-50CE.

7   God Most High will bless you and shine his face upon you, and he will open for you
8   his rich storehouse in the heavens, to send down upon your land
9   showers of blessing, dew and rain, the early rain and the latter rain in its season, and to give you frui[t],
10   produce, grain, wine and oil in abundance; and the land will produce for you [d]elightful fruit
11   so that you will eat and grow fat. *vac* And none will miscarry in your land,"

The Community Rule (Serkeh ha-Yahad)

Originally known as The Manual of Discipline, the Community Rule contains a set of regulations ordering the life of the members of the "yahad," the group within the Judean Desert sect who chose to live communally and whose members accepted strict rules of conduct. This fragment cites the admonitions and punishments to be imposed on violators of the rules, the method of joining the group, the relations between the members, their way of life, and their beliefs. The sect divided humanity between the righteous and the wicked and asserted that human nature and everything that happens in the world are irrevocably predestined. The scroll ends with songs of praise to God. A complete copy of the scroll, eleven columns in length, was found in Cave 1. Ten fragmentary copies were recovered in Cave 4, and a small section was found in Cave 5. The large number of manuscript copies attests to the importance of this text for the sect. This particular fragment is the longest of the versions of this text found in Cave 4.

"And according to his insight he shall admit him. In this way both his love and his hatred. No man shall argue or quarrel with the men of perdition. He shall keep his council in secrecy in the midst of the men of deceit and admonish with knowledge, truth and righteous commandment those of chosen conduct, each according to his spiritual quality and according to the norm of time. He shall guide them with knowledge and instruct them in the mysteries of wonder and truth in the midst of the members of the community, so that they shall behave decently with one another in all that has been revealed to them. That is the time for studying the Torah (lit. clearing the way) in the wilderness. He shall instruct them to do all that is required at that time, and to separate from all those who have not turned aside from all deceit. These are the norms of conduct for the Master in those times with respect to his loving and to his everlasting hating of the men of perdition in a spirit of secrecy. He shall leave to them property and wealth and earnings like a slave to his lord, (showing) humility before the one who rules over him. He shall be zealous concerning the Law and be prepared for the Day of Revenge. He shall perform the will [of God] in all his deeds and in all strength as He has commanded. He shall freely delight in all that befalls him, and shall desire nothing except God's will..."

Damascus Document (Brit Damesek)

The Damascus Document is a collection of rules and instructions reflecting the practices of a sectarian community. It includes two elements. The first is an admonition that implores the congregation to remain faithful to the covenant of those who retreated from Judea to the "Land of Damascus." The second lists statutes dealing with vows and oaths, the tribunal, witnesses and judges, purification of water, Sabbath laws, and ritual cleanliness. The right-hand margin is incomplete. The left-hand margin was sewn to another piece of parchment, as evidenced by the remaining stitches. In 1896, noted Talmud scholar and educator Solomon Schechter discovered sectarian compositions which later were found to be medieval versions of the Damascus Document. Schechter's find in a synagogue storeroom near Cairo, almost fifty years before the Qumran discoveries, may be regarded as the true starting point of modern scroll research.

4Q271(Df)

...with money...
...[his means did not] suffice to [return it to him] and the year [for redemption approaches?]...
...and may God release him? from his sins. Let not [ ] in one, for
it is an abomination....And concerning what he said (Lev. 25:14), ["When you sell anything to or buy anything from] your neighbor, you shall not defraud one another," this is the expli[cation...
...] everything that he knows that is found...
...and he knows that he is wronging him, whether it concerns man or beast. And if
[a man gives his daughter to another ma]n, let him disclose all her blemishes to him, lest he bring upon himself the judgement
[of the curse which is said (Deut. 27:18)] (of the one) that "makes the blind to wander out of the way." Moreover, he should not give her to one unfit for her, for
[that is Kila'yim, (plowing with) o]x and ass and wearing wool and linen together. Let no man bring
[a woman into the holy] who has had sexual experience, whether she had such experience
[in the home] of her father or as a widow who had intercourse after she was widowed. And any woman
[upon whom] there is a bad name in her maidenhood in her father's home, let no man take her, except
[upon examination] by reliable [women] who have clear knowledge, by command of the Supervisor over

[the Many. After]ward he may take her, and when he takes her he shall act in accordance with the law ...and he shall not tell...
[ ] L [ ]

Tongues of Fire (1Q29, 4Q376)

1Q29 F.1

(...) (...) the stone, just as the LORD commanded ....) and your Urim. And it (the cloud?) shall come forth with him, with the tongues of fire. The left-hand stone which is on its left side shall be uncovered before the whole congregation until the priest finishes speaking and after the cloud has been lifted ... And you shall keep (...) the prophet has spoken to you (...) (...) who counsels rebellion (...) (...) the LORD your God (...)
F.2
(...) (... the) right-hand stone when the priest comes out (...) three tongues of fire from the right-hand stone (...) (from ...) (...) and after he goes up he shall draw near to the people(...)
F.3-4
(...) (...the LORD) your God (...) (...Blessed is the God of Israel) (...) (...) among them all. Your name (...) (...and an) abundance of strenght, honored (and awesome...) (...)
F.5-7
(...) these words, according to all (...) (... and then) the priests shall interpret His will , all (...) (...) the congregation (...) (... O Children of Israel, keep all of these words) (...) (... to do) all (...) the number of commandments (...) (...) their (...)

4Q376

F.1 Col.1
 (...) the anointed priest upon whose head has been poured   the anointing oil ... and he shall offer a bull of the herd and a ram(...) for the Urim.
Col.2
and your Urim. And it (the cloud?) shall come forth with him, with tongues of fire. The left-hand stone which is upon its left side shall be uncovered before the whole congregation until the priests finishes speaking. And after the cloud has been lifted (...) And you shall keep (...) and the prophet has spoken to you.
Col.3
according to this entire commandment. And if the Leader of the whole nation is in the camp or (if ...) his enemy and Israel with him, or if they march on a city to throw up a siege against it, or in respect to any matter which (...) to the Leader (...) the field is far (...)

The Copper Scroll (3QTreasurea)

One of the most illusive documents found in the Qumran region is *The Copper Scroll*. Made of two separate sheets of copper, rolled up and oxidized right through, the contents of *The Copper Scroll* could only be determined after it had been cut into parallel strips. The text is difficult to read because it is virtually impossible to differentiate between some letters and others that are almost like them. The copyist made numerous mistakes thus making the task of the translators even more difficult. The document is mysterious. Is it legend from folklore about fictitious treasures or a catalogue of hiding places for real treasures? The formulas and directions are ambiguous and inconclusive thereby hinting at the possibility that the scroll is a myth. Furthermore, scholars presume that The Copper Scroll was written about 40 years after all the other scrolls. Specific and blatant contradictions among the translators forced us students to make educated guesses between the possible choices without certainty of the accuracy. For example, one translator suggested that the location of a treasure was facing a certain direction. Meanwhile another translator suggested that the entrance of the location is facing that direction, but location itself was facing in a different direction. Some treasure had a numeric value and other descriptions of the same treasure did not. Sometimes the treasure was gold, and other times it was silver. All together these examples combined to make the translated text ambiguous and intimate towards the fictional nature of the content.

Column I
In the ruin of Horebbah which is in the valley of Achor, under the steps heading eastward about forty feet: lies a chest of silver that weighs seventeen talents (yard stick). In the tomb of the third section of stones there is one hundred gold bars. Nine hundred talents are concealed by sediment towards the upper opening, at the bottom of the big cistern in the courtyard of the peristyle. Priests garments and flasks that were given as vows are buried in the hill of Kohlit. This is all of the votive offerings of the seventh treasure. The second tenth is impure. The opening is at the edge of the canal on its northern side six cubits toward the immersed pool. Enter into the hole of the waterproofed Reservoir of Manos, descend to the left, forty talents of silver lie three cubits from the bottom.

Column II
Forty two talents lie under the stairs in the salt pit. Sixty five bars of gold lie on the third terrace in the cave of the old Washers House. Seventy telents of silver are enclosed in wooden vessel that are in the cistern of a burial chamber in Matia's courtyard. Fifteen cubits from the front of the eastern gates, lies a cistern. The ten talents lie in the canal of the cistern. Six silver bars are located at the sharp edge of the rock which is under the eastern wall in the cistern. The cistern's entrance is under the large paving stone threshold. Dig down four cubits in the northern corner of the pool that is east of Kohlit. There will be twenty two talents of silver coins.

Column III
Dig down nine cubits into the southern corner of the courtyard. There will be silver and gold vessels given as offerings, bowls, cups, sprinkling basins, libation tubes, and pitchers. All together they will total six hundred nine pieces. Dig down sixteen cubits under the eastern corner to find forty talents of silver. Votive vessels and priestly garments are at the northern end of the dry well located in Milham. The entrance is underneath the western corner. Thirteen talents of silver coins are located three cubits beneath a trap door in the tomb in the north-east end of Milham.

Column IV
Fourteen talents of silver can be found in the pillar on the northern side of the big cistern in Kohlit. **SK** When you go forty-one cubits into the canal that comes from...you will find fifty-five talents of silver. Dig down three cubits in the middle of the two boulders in the Valley of Achor, and you will find two pots full of silver coins. At the mouth of the underground cavity in Aslah sit two hundred talents of silver. Seventy talents of silver are located in the eastern tunnel which is to the north of Kohlit. Dig for only one cubit into the memorial mound of stones in the valley of Sekaka to find twelve talents of silver.

Column V
A water conduit is located on the northern side of Sekaka. Dig down three cubits under the large stone at the head of this water conduit to discover seven talents of silver. Vessels of offering can be found in the fissure of Sekaka, which is on the eastern side of the reservoir of Solomon. Twenty-three talents of silver are buried quite nearby above Solomon's Canal. To locate the exact spot, go sixty cubits toward the great stone, and dig down for three cubits. Thirty two talents of silver can be located by digging seven cubits under the tomb in the dried up riverbed of Kepah, which is between Jericho and Sekaka.

Column VI
Forty-two talents of silver lie underneath a scroll in an urn. To locate the urn, dig down three cubits into the northern opening of the cave of the pillar that has two entrances and faces east. Twenty-one talents of silver can be found by digging nine cubits beneath the entrance of the eastward-looking cave at the base of the large stone. Twenty-seven talents of silver can be found by digging twelve cubits into the western side of the Queen's Mausoleum. Dig nine cubits into the burial mound of stones located at the Ford of the High Priest to find twenty-two talents of silver.

Column VII
To find four hundred talents of silver measure out twenty-four cubits from the water conduit of Q...of the northern reservoir with four sides. Dig six cubits into the cave that is nearby Bet Ha-Qos to locate six bars of silver. Dig seven cubits down under the eastern corner of the citadel of Doq to find twenty-two talents of silver. Dig three cubits by the row of stones at the mouth of the Kozibah river to obtain sixty talents of silver, and two talents of gold.

Column VIII
A bar of silver, ten vessels of offering, and ten books are in the aqueduct on the road that is to the east of Bet Ahsor, which is east of Ahzor. Dig down seventeen cubits beneath the stone that lies in the middle of the sheep pen located in the outer valley to find seventeen talents of silver and gold. Dig three cubits under the burial mound of stones located at the mouth of the Potter ravine to find four talents of silver. Dig twenty-four cubits below the northward burial chamber that is located on the south-west side of the fallow field of the valley of ha-Shov to reveal sixty-six talents. Dig eleven cubits at the landmark in the irrigated land of ha-Shov and you will find seventy talents of silver.

Column IX
Measure out thirteen cubits from the small opening at the edge of Nataf, and then dig down seven cubits there. Seven talents of silver and four stater coins lie there. Dig down eight cubits into the eastern-looking cellar of the second estate of Chasa to obtain twenty-three and a half talents of silver. Dig sixteen cubits into the narrow, seaward-facing part of the underground chambers of Horon to discover twenty-two talents of silver. A sacred offering worth one mina of silver is located at the pass. Dig down seven cubits at the edge of the conduit on the eastern side inside the waterfall to locate nine talents of silver.

Column X
When going down to the second floor, look to the small opening to find nine talents of silver coins. Twelve talents lie at the foot of the water wheel of the dried up irrigation ditches which would be fed by the great canal. Sixty-two talents of silver can be found by going to the left for ten paces at the reservoir which is in Beth Hakerem. Three hundred talents of gold and twenty penalty fees can be found at the entrance to the pond of the valley Zok. The entrance is on the western side by the black stone that is held in place by two supports. Eight talents of silver can be found by digging under the western side of Absalom's Memorial. Seventeen talents are located beneath the water outlet in the base of the latrines. Gold and vessels of offering are in this pool at its four angles.

Column XI
Very near there, under the southern corner of the portico in Zadok's tomb, beneath the pillars of the covered hall are ten vessels of offering of pine resin, and an offering of senna.
 Gold coins and consecrated offerings are located under the great closing stone that is by the edge, next to the pillars that are near by the throne, and toward the tip of the rock to the west of the garden of Zadok. Forty talents of silver are buried in the grave that is under the colonnades. Fourteen votive vessels possibly of pine and resin are in the tomb of the common people and Jericho. Vessels of offering of aloes and tithe of white pine are located at Beth Esdatain, in the reservoir at the entrance of the small pool. Over nine-hundred talents of silver are next to the reservoir at the brook that runs near the western entrance of the sepulchre room.

Column XII

Five talents of gold and sixty more talent are under the black stone at the Western entrance. Forty-two talents of silver coin are in the proximity of the black stone at the threshold at the sepulchral chamber. Sixty talents of silver and vessels are in a chest that is under the stairs of the upper tunnel on Mount Garizim. Six-hundred talents of silver and gold lie in the spring of Beth-Sham. Treasure weighing seventy-one talents and twenty minas are in the big underground pipe of the burial chamber at the point where it joins the house of the burial chamber. A copy of this inventory list, its explanation and the measurements and details of every hidden item are in the dry underground cavity that is in the smooth rock north of Kohlit. Its opening is towards the north with the tombs at its mouth.

Prayer for King Jonathan (Tefillah li-Shlomo shel Yonatan ha-Melekh, 4Q448)

The King Jonathan mentioned in this text can be none other than Alexander Jannaeus, a monarch of the Hasmonean dynasty who ruled Judea from 103 to 76 B.C.E. The discovery of a prayer for the welfare of a Hasmonean king among the Qumran texts is unexpected because the community may have vehemently opposed the Hasmoneans. They even may have settled in the remote desert to avoid contact with the Hasmonean authorities and priesthood. If this is indeed a composition that clashes with Qumran views, it is a single occurrence among 600 non-biblical manuscripts. However, scholars are exploring the possibility that Jonathan-Jannaeus, unlike the other Hasmonean rulers, was favored by the Dead Sea community, at least during certain periods, and may explain the prayer's inclusion in the Dead Sea materials. This text is unique in that it can be clearly dated to the rule of King Jonathan. Three columns of script are preserved, one on the top and two below. The upper column (A) and the lower left (C) column are incomplete. Differences between the script of Column A and that of B and C could indicate that this manuscript is not the work of a single scribe. This small manuscript contains two distinct parts. The first, column A, presents fragments of a psalm of praise to God. The second, columns B and C, bear a prayer for the welfare of King Jonathan and his kingdom. In column A lines 8-10 are similar to a verse in Psalm 154, preserved in the Psalms Scroll (11QPsa) exhibited here. This hymn, which was not included in the biblical Book of Psalms, is familiar, however, from the tenth-century Syriac Psalter.

Column A

1. Praise the Lord, a Psalm [of
2. You loved as a fa[ther(?)
3. you ruled over [
4. vacat [
5. and your foes were afraid (or: will fear) [
6. ...the heaven [

7. and to the depths of the sea [
8. and upon those who glorify him [
9. the humble from the hand of adversaries [
10. Zion for his habitation, ch[ooses

    Column C   Column B

1. because you love Isr[ael   1. holy city
2. in the day and until evening [2. for king Jonathan
3. to approach, to be [3. and all the congregation of your people
4. Remember them for blessing [4. Israel
5. on your name, which is called [5. who are in the four
6. kingdom to be blessed [6. winds of heaven
7.]for the day of war [7. peace be (for) all
8. to King Jonathan [8. and upon your kingdom
9.   9. your name be blessed

Hymns, Psalms and Poetry

Fernando Klein

The Chariots of Glory (4Q286-287)

We call this text, which contains some of the most beautiful and emotive vocabulary in the entire Qumran repertoire, the Chariots of Glory to emphasize its connections with Ezekiel's visions and Merkabah mysticism. It is a work of such dazzling faith and ecstatic vision that it fairly overwhelms the reader. Of course, it completely gainsays anyone who would challenge the literary audacity, virtuosity and creativity of those responsible for the Qumran corpus.

This work, which has obvious affinities with the already published Songs of the Sabbath Sacrifice, found at both Qumran and Masada, is a work of what goes by the name in Judaism and Kabbalah of the Mysticism of the Heavenly 'Chariot' or 'Throne' - so cultivated in the Middle Ages and beyond. If it is not the starting point of this genre, it is certainly one of the earliest extant exemplars of it.

Text

Foundations of Fire

Manuscript A

Fragment 1

(1) the seat of Your Honor and the footstools of the feet of Your Glory, in the Heights of Your standing and the ru[ng] (2) of Your Holiness and the chariots of Your Glory with their [mu]ltitudes and wheel-angels, and all [Your] Secrets, (3) Foundations of fire, flames of Your lamp, Splendors of honor, fi[re]s of lights and miraculous brilliances, (4) [hon]or and virtue and highness of Glory, holy Secret and pla[ce of Splendor and the highness of the beauty of the Fountain], (5) [majes]ty and the gathering-place of power, honor, praise and mighty wonders and healing[s], (6) and miraculous works, Secret Wisdom and image of Knowledge and Fountain of Understanding, Fountain of Discovery (7) and counsel of Holiness and Secret Truth, treasurehouse of Understanding from the sons of Righteousness, and dwelling places of Upright[ness...] (8) Pious O[nes] and congregation of Goodness and Pious Ones of Truth and Eternal Merciful Ones and miraculous Myst[eries] (9) when th[ey app]ear, and weeks of Holiness in their rightful order and monthly flags... (10) in their seasons and festivals of Glory in [their] times... (11) and sabbaths of the earth in their divi[sions and appointed] times of jub[ilee...] (12) [and] Eternal [Jub]ilees... (13) [Li]ght and Dark[ness...]

Ecology Hymn

Fragment 2

(1) (Let us praise)... the land and all who [dwe]ll... who inhabit it, earth and all the[ir] equipment (2) [and al]l its subsistance... [and al]l hills, valleys and all

streams, land of beau[ty...] (3) [Let us] praise the dept[hs] of forests and the wildernesses of Hor[eb...] (4) and its wilds and foundations of..., islands and... (5) [the]ir fruits, highland woods and all the cedars of Leba[non...] (6) [new] wine and oil and all the produce of... (7) and all the offerings of the land in (the) tw[elve] months... (8) Your word. Amen. Amen. (9)... and fortress and water, deep wells... (10) every stream, de[ep] rivers... (11) water... (12) [al]l their Secrets...

Eternal Knowledge

Fragment 3

Column 1

(1)... the lands (2)... their young men (3)... and all their companions in praises of (4)... and praises of Truth in the times of fe[stival] (5)... Your... and the bearer of Your Kingdom in the midst of Peo[ples] (6)... Angels of purity with all Eternal Knowledge, to... (7) [to bles]s Your glorious Name for[ever.] Amen. Amen. (8)... continue to praise the *God* of... [al]l His Truth...

The Community Council Curses Belial

Fragment 3

Column 2

(1) The Community Council shall say together in unison, 'Amen. Amen.' Then [they] shall curse Belial (2) and all his guilty lot, and they shall answer and say, 'Cursed be [Be]lial in his devilish (Mastematic) scheme, (3) and damned be he in his guilty rule. Cursed be all the spir[its of] his Mot in their Evil scheme. (4) And may they be damned in the schemes of their [un]clean pollution. Surely [they are the to]t of Darkness. Their punishment (5) will be the eternal Pit. Amen. Amen. And cursed be the Evi[l] One [in all] of his dominions, and damned be (6) all the sons of Bel[ial] in all their times of service until their consummation [forever. Amen. Amen.'] (7) And [they are to repeat and say, 'Cursed be you, Angel of the Pit and Spir[it of Destruction in al[l] the schemes of [your] gu[ilty] inclination, (8) [and in all the abominable [purposes] and counsel of [your] Wick[edness. And damned be you in [your] [sinful] d[omi]n[ion] (9) [and in your wicked and guilty rule,] together with all the abom[inations of She]ol and [the reproach of the P]it, (10) [and with the humiliations of destruction, with [no remnant and no forgiveness, in the fury of [*God*'s] wrath [for]ever [and ever.] Amen. A[men.] (11) [And cursed be al]l who perform their [Evil schemes,] who establish your Evil purposes [in their hearts against] (12) Go[d's Covenant,] so as to [reject the words of those who see] his [Tru]th, and exchange the Judge[ments of the Torah...]

The Splendour of the Spirits

Manuscript B

Fragment 1

(1)... as the[ir] teachers... (2) their... their honor... (3) their Glory, the doors of their wonders... (4) the Angels of fire and the Spirits of cloud... (5) [the] embroidered [Splen]dor of the Spirits of the Holy of Hol[ies...] (6) and firmaments of the Holy [of Holies...] (7) months with all [their] festivals... (8) the Glorious Name of Yo[ur] *God*... (9) and all the servants of Ho[liness...] (10) in the Perfection of th[eir] works... (11) in [their] wond[rous] Temples... (12) [a]ll [their] servant[s...] (13) Your Holiness in the habitat[ion of...]

Fragment 2

(1)... them, and they shall bless Your Holy Name with blessing[s...] (2) and they shall bless] You, all creatures of flesh in unison, whom [You] have creat[ed...] (3) beasts and birds and reptiles and the fish of the seas, and all... (4) [Y]ou have created them all anew... Fragment 3 (13)... The Holy Spirit [sett]led upon His Messiah...

Hymns of the Poor (4Q434, 436)

These texts are appropriately titled. It is important to see the extent to which the terminology Ebionim ('the Poor') and its synonyms penetrated Qumran literature. Early commentators were aware of the significance of this usage, though later ones have been mostly insensitive to it. The use of this terminology, and its ideological parallels, 'Am ('Meek') and Dal ('Downtrodden'), as interchangeable terms of self-designation at Qumran, is of the utmost importance.

Text

Fragment 1

(1)... Understanding to strengthen the downcast heart, and to triumph over the spirit in it; to comfort the Downtrodden in the time of their distress, and as for the hands of the fallen, (2) to hold them up so they can make vessels of Knowledge, to give Knowledge to the Wise, to increase the learning of the Upright, so as to comprehend (3) Your wonders that You performed in former days, in previous generations, the Eternal Insight that (4) You [established] before (establishing) me. And You kept your Law before me, and Your Covenant You confirmed for me, strengthening (it) upon [my] heart. (5)... to walk in Your Ways. You commanded my heart, and instructed my conscience not to forget Your Laws. (6)... You have... Your Law and opened my conscience, and strengthened me to pursue the Ways of

(7) [Truth...] Your... You made my mouth like a sharp sword and opened my tongue to words of Holiness. You have put (8)... instruction. Let them not meditate upon the doings of man, whose lips are in the Pit. You strengthened my legs, (9)... And by Your hand you strengthened (me) with days. You sent me in... (10) (a heart of stone?) You [re]moved from me and put a pure heart in its place, remov[ing] evil inclination

Fragment 2

Column 1

(1) Bless the Lord, 0 my soul, because of all His wonders forever. Blessed be His name, for He saved the soul of the Poor One (Ebion). (2) He has not despised the Meek ('Ani), nor has he forgotten the distress of the Downtrodden (Dal). (On the contrary), He opened His eyes to the Downtrodden, and, inclining His ears, hearkened to (3) the cry of orphans. In His abundant Mercy He comforted the Meek, and opened their eyes to behold His ways, and their ears, to hear (4) His teaching. And He circumcised the foreskin of their hearts, and saved them because of his Grace, and He directed their foot to the Way. He did not abandon them in their great distress, nor (5) give them into the hand of Violent Ones, nor judge them with the Wicked, nor kindle his wrath against them, nor destroy them (6) in His anger, though the wrath of His hot anger did not abate at all. But He did not judge them in fiery zeal; (7) (rather) He judged them in the abundance of His Mercy. The Judgements of His eye were to test them. In the greatness of His Mercy, He brought them (from) among the Gentiles; from the hands of (8) Man He saved them. He did not judge them in the multitude of the nations, nor scatter them among the Peoples. (Rather), He hid them in the shadow of His wings, (9) and made the dark places Light before them, and the crooked places straight, and He revealed to them abundant Peace and Truth. He made (10) their Spirit by measure, and meted out their words by weight, causing them to sing like flutes. He gave them a heart of rejoicing, and they walked in the Way of His heart. (11) But also, in the way of His heart He led them, because they... , their Spirit at rest, and raised up a testi[mony...] He commanded a plague to..., (12) And He set his Angel around the s[ons of Israel lest they be destroyed [in the land of] (13) their enemies... His wrath, to bring His anger... on them... (14) He hated... His Glory to...

Column 2

(1) in Evil... distress... (2) their works... for them against the sons of Man, and You saved them for Your own sake... (3) And they aggravated their iniquity and the iniquity of their fathers, but they atoned for it in... (4) Judgements, and to the Way that... (5) again, because... their... in...

Fragment 3

(1) Your... to be comforted on account of her mourning; her affliction He... (2) nations [He will destroy, and peoples cut off, and the Wicked... He fashioned (3) the works of Heaven and earth, and they met, and His Glory filled... their [Tr]uth (4) will make atonement. Goodness will multiply, and the Goodness of the... will comfort them .... to eat (5) its fruit and its Goodness. (6) Like a man whose mother comforts him, so will He comfort them in Jerusal[em... Like a bridegroom] over the bride, over her (7) . . and He will put... and He will lift a]p His Throne forever and ever. And His Glory... and all the Gentiles (8)... and there will be... desire (9)... forever the radiance of... I shall bless (10) [the Name]... blessed be the Name of the Most High... (11)... Your Piety (or 'Grace') upon him (12)... for the sake of the Torah You established (13)... the book of Your Laws...

Thanksgiving Hymns (1QH)

Lawrence H. Schiffman wrote that it is tempting to regard the Thanksgiving Scroll as a series of hymns for public worship. This is what I perceived them to be at first. After researching it further, I discovered that there has been no evidence discovered that concludes or supports this. The passages were probably written by the leader of a sect or church. Some scholars believe that it was written by the Teacher of the Righteousness. The poems belong to a devotion of some sort by the people for their God. This can be cross referenced with the Songs of the Sabbath; where God in the heavens was praised daily according to fixed rituals. Later it was discovered that it was not a praise of God by people on earth, but the angles' praise of God in heaven. After further exploration, I found that the Hymns of Thanksgiving could have been used to have the spiritual effect of the individual members of the community. Penetrating them and therefore dominating their spirit as if being controlled by a force. The author of these passages, assumed to be the Teacher of Righteousness, gives a vivid description of mankind being 'in sin,' the constant struggle with the forces of evil around him, mankind having bad temptations from birth to death.

Another interesting point in the 1st fragment is the use of the word Belial. This word was substituted for the Angel of Darkness. Is Belial conceived to be a real person in the Hymns or is the author speaking in general terms of mankind? This use of the word Belial or Angel of Darkness also refers to the Teacher of Righteousness who uses this word in his frequent works, (1QH 2:16, 4:10, 4:13, 5:26, 5:39, 6:21, 7:3).

Hymn #7 (formerly 2):

I give thanks to you, Lord, because you have placed me in the circle of life and guided me against the evils of the world. Violent men have threatened my life because of my faith in you Lord. For they are an assembly of trickery and a crowd of evil, the do not know that through you I live and in your compassion you will spare me in my soul. Because of you they raid my life to spite you by the judgement of the wicked. But you give me strength in the faces of the unworthy. And I said, mighty men have pitched their camps and swarmed against me with all the temptations of unjustly things. They have begun things which have no cure, no stopping. Their weapons of evil engulf the land like a tidal wave upon the shore. Like a wave of destruction devouring a multitude of men. Temptation rose inside me but my soul clung to the faith of you Lord. They have fallen to the destruction of each other which they brought on themselves, but I will not fall to the rein of their destruction, for I keep upon level ground and apart from them I will bless you Lord.

Hymn #8 (formerly 3):

I give thanks to you, Lord, for you have [fastened] your eye upon me. You have save me from the passion of lying deception, and from the congregation of those who seek wealth. You have blessed the soul of the poor one who planned to destroy me by spilling my blood while I was at service to you. But they did not know that my soul belonged to you, so they made a mockery of me in the mouths of all that seek for lies. [...] But you, my Lord, have restored the faith of the poor and the needy against one stronger than me; you have saved my soul from the hand of the mighty. You have not permitted their insults to pursue me into craving their service.

The Thanksgiving Psalms

Psalm 4.

I thank you, O Lord,
for your eye is awake and watches over my soul.
You rescue me from the jealousy of liars,
from the congregation of those who seek the smooth way.
But you save the soul of the poor
whom they planned to destroy
by spilling the blood of your servant.
I walked because of you - but they didn't know this.
They laughed at me. They shamed me
with lies from their mouth.
But you helped the soul of the poor and the weak,
you saved me from their harsh arms,
you redeemed me amid their taunts.
From the wicked I do not fear destruction.

Fernando Klein

Psalm 5.

They made my life a ship on the deep sea,
like a fortified city circled by aggressors.
I hurt like a woman in labor bearing her first child,
whose belly pangs torture her in the crucible.
Pains of Hell
for a son come on the waves of death.
She labors to bear a man,
and among the waves of death she gives birth to a manchild,
with pains of Hell.
He springs from the crucible,
O wondrous counselor with power :
Yes, a man emerges from the waves..
But she who carries dead seed in her womb
suffers waves from a pit of horror.
The foundations of the wall will rock
like a ship on the face of the waters.
Clouds will bellow.
Those who dwell in the dust, like those on the sea,
are terrified by the roar of the waters.
All those wise men are like mariners on the deep:
their wisdom confounded by the roaring seas.
The abyss boils over the fountains of water.
The seas rage.
Hell opens, and arrows fly toward Heaven.
Their eternal bars are bolted.

Psalm 8.

I thank you, O Lord.
You illumined my face by your covenant.
I seek you,
As sure as the dawn you appear as perfect light.
Teachers of lies have comforted your people
and now they stumble, foolishly.
They abhor themselves
and do not esteem me through whom your wonders
and powers are manifest.
They have banished me from my land like a bird
from its nest, and my friends
and neighbors are driven from me.
They think me a broken pot.
They preach lies. They are dissembling prophets.
They devise baseness against me,
exchanging your teaching, written in my heart,

for smooth words.
They deny knowledge to the thirsty
and force them to drink vinegar to cover up error.
They stumble through mad feasts,
but you, God, spurn the schemes of Belial.
Your wisdom prevails.
Your hearts meditation prevails, established forever.

Psalm 23.

Your holy spirit
illuminates the dark places of the heart
of your servant,
with light like the sun.
I look to the covenants made by men,
worthless.
Only your truth shines,
and those who love it are wise
and walk in the glow
of your light.
From darkness you raise hearts.
Let light shine on your servant.
Your light is everlasting.

Apocryphal Psalms (4QPsf=4Q88, 4QapPs=4Q448, 11QPsa-b=11Q5-6)

The following sections of the Dead Sea Scrolls are commonly referred to as the Apocryphal Psalms or the Apocryphal Psalms of David. The scrolls were found in the caves at Qumran, along with many others. These specific scrolls were discovered in caves 4 and 11. These scrolls tell of the great deeds of God and of David, as they praise the works and actions of both. Included in the scrolls containing the Apocryphal Psalms of David are Psalms 151 and 154, which are not normally found in the Bible, in which the book of Psalms ends with Psalm 150. The scrolls can be found in books containing Texts of the Dead Sea Scrolls, in Bibles which include the Apocrypha, and in some books containing songs and prayers which are not usually found in a conventional Bible. Much of the contents of the caves were discovered by bedouins or nomadic people. However, the majority of the excavation of those same caves was handled by professionals, trained to take special care of the precious contents of these ancient caverns. Many of the scrolls which will be explained herein are only partial, due to the deterioration caused by time and neglect. As a result of this, the reader is left with fragments of the original texts. But, luckily, much of what remained of the scrolls was still readable for a small group of highly trained scholars. The contents of cave 4 were discovered in 1952. This cave is commonly regarded to be the central library of the Qumran community. The find included 15,000 fragments which came from roughly 550 different manuscripts. Cave 11 was found to contain manuscripts in January 1956. The contents of both of these caves proved to be quite lucrative for the people who found these scrolls and cared for them before their eventual sales to both museums and private collectors.

A scroll, found in Cave 11, commonly known as 11Q6 Column 19, was one of the longer pieces which was found at Qumran. Its surface was the thickest of any of the scrolls because it is possible that it was written on calfskin, rather than on sheepskin as were the majority of other scrolls. While the script of this particular scroll was of very fine quality, several of the lines of the bottom of the scroll were missing.

4Q448

Column XVIII

Psalm 154

Praise God in a loud voice. Testify to his glory in the assembly.
Lift up His name with the righteous and speak of His greatness with the faithful.
Become one with the perfect and the good to praise the Lord.
Join and worship together to tell of His salvation. And be swift in making known His fortitude and His righteousness to all the simple.
Knowledge is granted so that we may praise the Lord and tell of his greatness.
She is made known to mankind, to speak of His strength and tell of his greatness to the ignorant, who have strayed from her gates and have sinned.
For God is the Lord of Jacob and his grace is seen in all his works.
A person who praises God is recognized by Him because the worshiper brings and offering and sacrifice of livestock,   because the worshiper fills the altar with gifts.
Her voice and her songs are heard and sung by the righteous.
When they feast together, she is mentioned.
Their thoughts are on the Law of God and they speak to testify of His strength.
The evil and the rebellious are far from her grace.
See how God has mercy on the good, and it is great for those who praise Him; He is their soul's salvation from wickedness.
Praise the Lord who saves the meek from the grasp of the unknown and delivers the righteous from evil,
Who lifts up a horn from Jacob and a moderator from Israel.
He wants his gathering place to be in Zion, and He picks Jerusalem for all eternity.

4Q88

Column IX
[. . .]

The Dead Sea Jesus

The masses will worship God because He has come to judge everything and to rid the earth of evil, so that sinners shall find no repose, the heavens shall give their due, and there will be no wrong doings there.
The earth will produce crops in its season and they shall not fail.
The fruit trees shall [. . .] of their vineyards and their springs will not dry up.
The poor will eat for those who follow YHWH shall not go hungry.

Column X
[. . .]
[. . .] meanwhile the heavens and earth will praise together
And all the evening starts will then adore.
Rejoice, Judah, be happy!
Be glad and let your joy shine forth!
Keep your feasts and your oaths because within you there is no Belial -- biblical name of the devil --.
Raise your hand, make your right hand strong. See, your foes shall be eliminated.
And all evil doers shall fell.
But you, YHWH, shall remain forever.
Your glory is everlasting.
[Hallelujah!]

11Q5

Column XXVIII

Hallelujah. Of David, son of Jesse.
Psalm 151
My brothers were bigger than me and I was the youngest of my father's sons; He made me the master of His flocks, and shepherd of his goats. My hands created a flute, my fingers a lyre, and I praised God. I told myself that neither the mountains nor the hills tell me of the glory of God, nor the trees His words, nor the flocks His actions.
Who, then will tell of God's deeds? God saw all
He heard all and listened. He sent his prophet to anoint me, Samuel, to sift me up. My brothers went out to meet him, well built, beautifully presented. They were very tall
and had lovely hair, but the Lord did not pick them, He sent for me from tending the flock
and anointed me with holy oil and made me a ruler of His people and of the sons of His covenant.
[. . .]
First of David's exploits after the Lord's prophet had anointed him
Then I saw a Philistine threatening from enemy lines [. . .]

11Q6

Column XIX

Fragment A
Impoverished and feeble am I for [. . .]
For not even a worm can praise You nor insects recount Your grace
The living can thank You and those who fall shall praise You highly.
You show them the ways of Your holiness and grace for You care for the souls of every living thing;
You provide for all living things. Judge us, O God, with Your kind ways, Your grace, and
Your Justice. The Lord hears the please of his followers.
He has not shunned them. Praise be to God who does good things and rewards his followers with His kindness. May my spirit lift up Your Name, to recount with joy Your righteous deeds, and proclaim Your eternal steadfastness.

Fragment B

And in Your grace, I have sought sanctuary. The images of Your might life up my heart.
I find peace in Your righteousness. Forgive me, Lord, and free me from my sins.
Give to me a sense of honour and knowledge. May I not be shamed in ruin.
Protect me from unclean spirits, save me from pain and temptation.
For You, O Lord, are my salvation, and I praise You everyday.
My people rejoice with me and are awestruck by Your power.
I will adore You and worship You for all eternity.

Plea for Deliverance (11QPsa=11Q5 col.xix)

Some Scholars have argued that 11QPsa should be considered as a canonical and therefore authoritative, open ended canon of psalms. The completion of the Masoretic psalter as canon formed an important step in that process. Thus, researchers place the completion of the Masoretic psalter as canon at the end of the first Century in Jamnia.

On the other hand, some people has opposed this view arguing the reverse, i.e. that 11QPsa is merely a liturgical collection with no real authority at all and no bearing on the Masoretic psalter as canon, which at least in its first four books was complete by the fourth century BC, and the final section not much later. There is no real evidence that 11QPsa was dependent on Masoretic but it could be dependent on a common tradition of psalm materials. Furthermore both the Masoretic psalter and 11QPsa seem to function as liturgical collections.

"Surely a maggot cannot praise thee nor a grave worm recount thy loving-kindness. But the living can praise thee, even those who stumble can laud thee. In revealing thy kindness to them and by thy righteousness thou dost enlighten them. For in thy hand is the soul of every living thing; the breath of all flesh hast thou given. Deal with us, O LORD, according to thy goodness, according to thy great mercy, and according to thy many righteous deeds. The LORD has heeded the voice of those who love his name and has not deprived them of his loving-kindness. Blessed be the LORD, who executes righteous deeds, crowning his saints with loving-kindness and mercy. My soul cries out to praise thy name, to sing high praises for thy loving deeds, to proclaim thy faithfulness--of praise of thee there is no end. Near death was I for my sins, and my iniquities have sold me to the grave; but thou didst save me, O LORD, according to thy great mercy, and according to thy many righteous deeds. Indeed have I loved thy name, and in thy protection have I found refuge. When I remember thy might my heart is brave, and upon thy mercies do I lean. Forgive my sin, O LORD, and purify me from my iniquity. Vouchsafe me a spirit of faith and knowledge, and let me not be dishonored in ruin. Let not Satan rule over me, nor an unclean spirit; neither let pain nor the evil inclination take possession of my bones. For thou, O LORD, art my praise, and in thee do I hope all the day. Let my brothers rejoice with me and the house of my father, who are astonished by the graciousness... [ ] For e[ver] I will rejoice in thee."

The Children Of Salvation (Yesha') and the Mystery of Existence (4Q416, 418)

If we are justified in grouping all these fragments together, this is one of the longest extant manuscripts in the unpublished corpus. Strictly speaking, the work as a whole has the character of an admonition and belongs in Chapter 5, but because of its eschatological thrust, mysticism, emphasis on the Mysteries of *God* and parallels with the preceding works and Hymns from Cave 1, we have chosen to place it here.

Text

The Eternal Planting

Fragment 1

(1) Open your lips (as) a Fountain to bless the Holy Ones. O ye, bring forth praise as an Eternal Fountain... For He has separated you from all (2) bodily spirit. O ye, separate yourself from all that He hates, and keep yourself apart from all Abominations of... He made all (flesh), (3) and caused every man to inherit his portion. He set you apart-and your portion-among the sons of Adam... He gave you authority. O ye, (4) this was how He glorified it when you sanctified yourself to Him, when He made you a Holy of Holies... for all... (5) He decided your fate and greatly increased your Glory, and made you as a firstborn for Himself among...(6) 'and I will give you My Goodness.' O ye, is not His Goodness yours? So always walk in His Faithfulness in all of (7) your works. O ye, seek His judgements from every hand, and the abundance of... (8) love Him, for with everlasting Piety

(Hesed) and mercies on all the Keepers of His word, and... (9) O ye, He has [op]ened up insight for you, and given you authority over His storehouse, and the accurate value for a measure (ephah) He has determined... (10) they... you. It is in your power to turn aside wrath from the Men of His Favor, and to appoint... (11) with you, before you take your portion from the hands of the Glory of His Holy Ones, and in... (12) He opened... and all who are called by His Holy Name... (13) with all the Er[as of] His sub[lime] radiance for an Eter[nal] Planting... (14) all those who inherit the land will conduct themselves, for in... (15) O ye, because of the Wisdom of your hands, He has given you authority, and [your] Knowledge... (16) a storehouse (?) for all humanity. From there you will designate your unclean food, and... (17) Seek understanding with all (your) might, and from every hand, take increased insight... (18) Bring forth what you lack for all those seeking after (their own) desire(s). Then you will understand... (19) You will be filled and satiated with abundant Goodness, and by the skill of your hands... (20) Because *God* has apportioned the inheritance of eve[ry living being,] and all those wise of heart have considered...

The Fountain of Living Water

Fragment 2

(1)... farmers, until all... (2) bring in your baskets, and your storehouses, all... (3) and the plain, season by season seek them out, and do not cease... (4) all of them seek in their season(s), and according to Wis[dom,] a man... (5) like a Fountain of Living Water which all me[n...] (6) and it is a hybrid like a mule, or like clo[thing made of two materials...] (7) in cattle and in... and also, your produce will be... (8) your Riches with your flesh...

All the Eras of Eternity

Fragment 3

(1)... concerning... all... (2) together with him... all who... (3) and they feared the deep, and... (4) And every sacrifice are you to offer them perpetually, and the peace (offering)... (5) in all the Eras of Eternity, because He is (the) *God* of Truth... (6) making the Righteous discern Good from Evil... (7) because it is the inclination of the flesh, and those who understand...

The Foundations of the Universe Shout out judgement

Fragment 4

(1)... your breath (2)... and you will understand... death with (3)... Shall they not walk in Truth? (4)... and all their joys with Knowledge. O ye foolish of heart, what is Goodness without (5) [...And how] can there be peacefulness without destruction? And how can there be judgement without establishing it, and how the dead will groan on account of al[l...] (6) you... you were created, but your backsliding leads to eternal damnation, because you walk... (7) The dark places will be made Light because of Your abundance, and Eternal Being (shall be the lot of) the Seekers of Truth and the Witnesses of Yo[ur] Judgements. (8) All the foolish of heart will be destroyed, and the sons of Wickedness will be found no more, and all those seekers after Evil will be ab[ashed...] (9) The Foundations of the Universe will shout out Your Judgement, and all the... will thunder... all the lovers of... (10) You will be the Elect of Truth, and pursuers after [insight with] Judg[ment...] those who are watchful... (11) according to all Knowledge. How can you say, 'We have worked for insight and stayed awake pursuing Knowledge of... in all... (12) But He has not tired during all the years of Eternity. Does He not delight in Truth forever?... Knowledge ministers unto Him, and [all the Angels of] (13) Heaven-whose inheritance is eternal life-would they ever say, 'We have grown weary in the ministries of Truth, and tir[ed in...] (14) for all ages do they not walk in Eternal Light?... again, and abundant radiance dwells with them. You... (15) in the firmaments of... the Foundation (possibly 'Secret') of the Pillars, all... 0 ye, in...

Eternal Glory

Fragment 5

(1) [...No]t one from all of their host will rest... (2) in Truth from the hand of all the storehouses of men... (3) Truth, and the measure of Righteousness He meted out to all... (4) [...dis]tributing them in Truth. He put them in place, and sought out their pleasures... (5) and a shelter for all, nor shall they exist without His favor and... (6) Judgement to visit repentance upon the Lords of Evil and the visitation of... (7) and to shut before the Evil Ones, and to lift up the head of the Downtrodden... (8) in Eternal Glory and peace everlasting, and the Spirit of Life to sepa[rate...] (9) all the sons of Eve, and with the power of *God* and the abundance of His Glory, with His Goodness... (10) and in His Faithfulness, they shall prostrate themselves continually all the day and praise His Name, and... (11) O ye, walk in Truth with all the [See]kers... (12) For His storehouse is under your authority, and whoever seeks his own aim (must do so) from your basket, and (to) them... (13) And if He does not stretch out His hand for your needs, then will His storehouse (provide) this need... (14)... and he shall not provide for his own wishes, for He shall not... (15) your hand. He will increase you[r] cattle abundantly... (16) forever...

The Scales of Righteousness

Fragment 6

(1)... your Fountain. Nor will you find what you lack, and your soul will languish for want of all Goodness, even unto death...(2) [... will be trou]bled all the day, and your soul will yearn to come into her (Wisdom's?) gates, and a grave (?) and clo[thing...] (3) your... And it will be as food to eat and fuel for the flame against... (4) For by your conduct you have troubled those who se[ek] pleasure and also... (5)

for you... because *God* made the pleasures of (His) storehouse, and meted them out in Truth... (6) [Fo]r in the scales of Righteousness He weighed out all their understanding, and in Truth...

The Angels of *God*'s Holiness

Fragment 7

(3)... its Ways are carved in suffering. You calm... (4) and there will be Lying in the heart of all the sons [of Adam...] 'He will trust in all of My Ways... (5) Knowledge. They have not earnestly sought Under[standing, nor Knowledge] chosen. Does not *God* [give Knowledge of... (6) on Truth, to discern all [Mysteries, and Under]standing did He apportion to those who inherited Truth. (7)... Lying. In Tr[uth... all His works. Is not peace and tranquility... (8) [Do you not k]now? Have you not heard? Surely the Angels of Go[d]'s Holiness are in Heaven... (9) Truth. And they pursue all the roots of Understanding, and diligently... (10) [according to] their Knowledge, one man will be glorified over another, and according to his insight will his honor be magnified... (11) For a man murmurs because he is lazy, and if a son of Adam is silent, is it not (because)... (12) And they will inherit an Eternal possession. Have you not seen .?

The Mystery of Existence

Fragment 8

(1)... to the fearful. You shall teach the first... (2) in former [times], why it existed and what existed in... (3) why they were... existence in... (4) [day and] night meditate on the Mystery of Existence... (5) [al]l their Ways, with the commands... (6) concerning the Knowledge of the Secret of Truth... (7) suffering and dominion... (8) [to] walk in the inclination of [His] Un[derstanding...] (9) walk...

The Salvation (Yesha') of His Works

Fragment 9

Column 1

(1) [...the] time, lest he hear you. And while he is alive, speak to him, lest he... (2) without appropriate reproof for his sake. Is it not bound up... (3) Furthermore, his Spirit will not be swallowed (i.e. 'consumed'), because in silence... (4) [and] quickly take his reproof to heart, and be not proud because of your transgressions... (5) He is Righteous, like you, because he is a prince among... (6) He will do. For how is He unique? In all His work, He is without... (7) Do not consider the Evil Man as a co-worker, nor anyone who hates... (8) the Salvation (Yesha') of His works, together with His command; therefore know how to conduct yourself with

The Dead Sea Jesus

Him... (9) Do not remove [the Law of *God*] from your heart, and don't go very far along by yourself... (10) For what is smaller than a man without means? Also, do not rejoice when you should be mourning, lest you suffer in your life... (11) existence; therefore, take from the children of Salvation (Yesha'), and know who will inherit Glory, for it is necessary for Him, not... (12) And instead of their mourning, (yours will be) everlasting joy, and the troublemaker will be placed at your disposal, and there will not [be...] (13) To all your young girls, spea[k] your judgements like a Righteous ruler, do not... (14) and do not take your sins lightly. Then the radiance of... will be... Judgement... (15) will He take, and then *God* will see, and His anger will be assuaged and He will give help against [your] sins, according to... (16) will not stand up all of its days. He will justify by His judgement, and without forgiving your... (17) Poor One. O ye, if you lack food, your need and your surplus... (18) You should leave as sustenance for His flocks according to His will, and [fr]om it, take what is coming to you, but do not add there[to...] (19) And if you lack, do not... Riches from your needs, for [His] storehouse will not be lacking. [And upon] (20) His word everything is founded, so a[at] what He gives you, but do not add to... (21) your life... If you borrow Riches from men to fill your needs, do not... (22) day and night, and do not for the peace of your soul... He will cause you to return to... Do not lie (23) to him. Why should you bear (the) sin? Also, from reproach... to his neighbor. (24)... and he will close up his hand when you are in need. According to Wisdo[m...] (25) and if affliction befalls you, and... (26) He will reveal... (27) He will not make atonement with... (28) a[gain]. Furthermore...

Your Holy Spirit

Fragment 9

Column 2

Fragment 10

Column 1

(1) He opened His Mercies... all the needs of His storehouse, and gave sustenance (2) to every living thing. There is none... [If he] closes his hand, and the Spi[rit of all] (3) flesh is withdrawn, you shall not... in it, and [with] his reproa[ch] will your face be covered, but by your arbitration (4) [he will go forth] from prison, like... And if he receives a loan, he [repays] (it) quickly in full. O ye, recompense him, for your purse (5) of treasures belongs to the one you are obliged to, (even if only) for the sake of your neighbors. You will... all of your life with him. (Therefore), quickly return (him) whatever (6) belongs to him; otherwise he will take your purse. In your affairs, do not compromise your Spirit. Do not exchange your Holy Spirit for any Riches, (7) because no price is worth [your soul.] Willingly seek the face of him who has authority over your storehouse, and in his own tongue (8) [speak with him.] In that way will you find satisfaction... Do not forsake your

Laws, and keep (secret) your Mysteries. (9)... If He assigns His service to you... (don't allow) sleep (to enter) your eyes until you have done it (10) [all... d]o not add. If th[ey] are needy... and do not be generous to him. Also, Riches without (11)... your [eye] shall see because of the abundant zeal of (12)... By His will, devote yourself to His service, and the Wisdom of His storehouse (13)... you will advise him, and become for him a firstborn son, and he shall love you as a man loves his only child. (14) Because you... [O y]e, do not rely on that... and do not stay awake at night because of your money, (15) [and during the night], continue su[ffer]ing because of it. Furthermore, do not demean your soul on account of someone who is not worth it, but rather be (16) to him... Do not strike someone who does not have your strength, lest you stumble and greatly humiliate yourself. (17)... your soul with the Goodness of Riches. You will be tilling the wind, and will serve your lord in vain; so, (18) do not sell your Glory for money, and do not transfer it as your inheritance, lest your bodily heirs be impoverished. Do not promise them (19)... If there are no cups, do not drink wine, and if there is no food, do not request delicacies. O ye, (20) [...If you] lack bread, do not glory in your poverty. You are needy... (21) [Do not] plunder to stay alive, and also, do not water down (the contents of) a vessel... [yo]ur Laws...

All the Ways of Truth

Fragment 10

Column 2

(3) So remember that you are needy... what you want (4) you shall not find. In your unfaithfulness, you will... He has appointed for you. (5) Do not reach your hand out for it, lest you be burned, [and] your body be consumed in His fire like... Thus He repaid him. (6) But there will be joy for you if you purify yourself of it. Also, do not take Riches from a man you do not know, (7) lest it only add to your poverty. If (*God*) has ordained that you should die in [you]r poverty, so He has appointed it; but do not corrupt your Spirit (8) because of it. Then you shall lie down with the Truth, and your sinlessness will He clearly proclai[m to th]em (the recording Angels). As your destiny, you will inherit (9) [Eternal] bliss. [For] though you are Poor, do not long for anything except your own portion; and do not be swallowed up by desire, lest you backslide (10) because of it. And if He restores you, conduct yourself honorably. And inquire among His chidren about the Mystery of Existence; then you will gain Knowledge of (11) His inheritance and walk in Righteousness, for He will... Do not... in all your Ways. Do homage to those who give you Glory, (12) and praise His Name continually, because out of poverty has He lifted your head, seating you among nobles. (13) He has given you authority over an inheritance of Glory, so seek His favor continuously. Though you are Poor, do not say, 'I am penniless, so I cannot (14) seek out Knowledge.' (Rather,) bend your back to all discipline, and through al[l Wisdo]m, purify your heart, and in the abundance of your (15) intellectual potential, investigate the Mystery of Existence. And ponder all the Ways of Truth, and consider all the roots of Evil. (16) Then you

will know what is bitter for a man, and what is sweet for a person. Honor your father in your poverty (17) and your mother by your behavior. For a man's father is like his arms, and his mother is like his legs. Surely (18) they have guided you like a hand, and just as He has given them authority over you and appointed (them) over (your) Spirit, so should you serve them. And just as (19) He has opened your ears to the Mystery of Existence, (thus) should you honor them, for the sake of your own honor. Just as... revere them, (20) for the sake of your own life and to lengthen your days. Even though you are in poverty... (21) unlawfully. If you take a wife in your poverty, take her from among the daughter[s of...] (22) from the Mystery of Existence. In your companionship, go forward together. With the helpmate of your flesh...

Fernando Klein

Calendrical Texts and Priestly Courses

Priestly Courses I (4Q321)

The first part of Mishmarot B delineates the equivalences between the solar and the lunisolar calendars. It also preserves information on the 'astronomical observance' of the moon which apparently acted as a check on the tabulated lunar month. The observance ascertained whether or not the full moon was waning at the proper rate, normally confirming the calculation of the day on which that month would end, and, concomitantly, when the subsequent one would begin.

Fragment 1 preserves the equivalences starting with the seventh month of the first year and ending with the second month of the fourth year. Fragment 2 begins with the fifth month of the sixth year and completes the cycle. Between the two fragments we learn how the intercalation of the lunisolar calendar was carried out at the end of the third year (by the solar calendar's reckoning; according to the lunisolar reckoning, after the first month of the fourth) and at the end of the sixth year.

The remaining portions of Fragment 2 describe the six-year cycle of First Days (of the months) and festivals in terms of the course to which they fall.

Text

Fragment 1

Column 1

(1) [on the first of Jedaiah on the twelfth of it. (The next lunar month ends) on the second of Abi[jah, on] the twe[nty-fifth of the eighth (solar) month. Lunar observation takes place on the third (2) of Mijamin, on the seventeenth] of it (i.e., of solar month eight). (The next lunar month ends) on the third of Jakim ninth (solar) month. Lunar observation takes place on the fourth (3) of Shecaniah, on the eleventh of it. (The next lunar month ends) on the fifth day of Immer, on the twenty-third of the ten[th (solar) month. Lunar observation takes place on the sixth day of J]eshebeab, (4) [on the tenth of] it. (The next lunar month ends) on the [si]xth of Jehezkel, on the twenty-second of the eleventh (solar) month. [Lunar observation takes place on the sabbath of] Pethahiah, (5) [on the ninth of it.] (The next lunar month ends) on the first of joiarib, on the twenty-second of the twelfth (solar) month. [Lunar observation takes place on the second of Delaiah, (6) [on the ninth of it. The] secon[d] (year): The first lunar month ends on the seco[n]d of Malachiah, on the twentie[th of the first (solar) month.] Lunar observation takes place on (7) [the third of Harim, on the seventh] of it. (The next lunar month ends) on the fourth of Jeshua, on the twentieth of the second (solar) month. [Lunar observation takes place on the fifth of Hakkoz, (8) on the seventh of it. (The next lunar month ends) on the fifth of Huppah, on the nineteenth of the third (solar)

month. Lunar observation takes place on the sixth of [E]l[iashib,] on the si[xth of it. (The next lunar month ends) on the sabba]th of Happizzez,

Column 2

(1) [on the eighteenth of the the fourth (solar) month. Lunar observation takes place on the first of Immer, on the fifth] of it. (The next lunar month ends) on the first of [Gamul, on the seventeenth of the fifth (solar) month. (2) Lunar observation takes place on the second of Je]hezk[el, on the fourth of it. (The next lunar month ends) on the third of Jeda]iah, on the [seventeenth of the sixth month. Lunar observation takes place on the fourth of] (3) Maaziah, on the fourth of it. (The next lunar month ends) on the fou[rth of Mijamin, on the fifteenth] of the seventh month. Lunar observation takes place on the fifjth of Seorim, on the third] (4) of it. (The next lunar month ends) on the sixth of Shecaniah, on the fifteenth of the eighth (solar) month. Lunar observation takes place on the sabbath of Abijah, on the second of it. [(The next lunar month ends) on the sabbath of Bilgah,] (5) on the fourteenth of the ninth (solar) month. Lunar observation takes place on the [first of Huppah, on the first] of the ninth (solar) month. A second lun[ar observation] takes place on the third of [Hezir, on the thirty-] (6) first o[f it. (The next lunar month ends) on the] second of Pethahiah, on the this[teenth of the tenth (solar) month.] Lunar observation takes place on the fourt[h of ja]chin, on the ninete[enth of it.] (7) [(The next lunar month ends) on the third of Delai]ah, on the twelfth of the eleven[th (solar) month. Lunar observation takes place on the sixth of Joiar[ib,] on the twentynin[th of i]t. [(The next lunar month ends) on the fifth of Harim,] (8) on the twe[lf]th of the twelfth (solar) month. Lunar observ[ation] takes place on the sabbath [of] Mijamin, on the twenty-eighth, on the [twenty-fourth [of the of it. The third (year): (The next lunar month ends) on [the sixth of Hakkoz, on the tenth...]

Column 3

(3) [on the seventh of the fifth (solar) month. Lunar observat]ion takes place on the first of Ha[rim, on the twen[ty-I [fourth] of it. (The next lunar month ends) on the sabbath of Mala[chiah, on the seventh of the sixth (solar) month. Lunar observation takes place on the second of Hakkoz, on the twenty-third of it. (4) (The next lunar month ends) on the first of Jeshua, on the fifth] of the seventh (solar) month. Lunar observation takes place on the fourth of Eliashib, on the tw[enty]- second [of it. (The next lunar month ends) on the third of Huppah, on the fifth of the eighth (solar) month. Lunar observation takes place on the fifth (5) of Bilgah, on the twenty-first in it. (The next lunar month ends) on the fourth of Hezir, on] the fourth of the ninth (solar) month. Lunar observation takes place on the sabbath of jeh[ezkel, on the twenty-first of it. (The next lunar month ends) on the sixth of Jachin, on the third of the tenth (solar) month. (6) Lunar observation takes place on the first of Maaziah, on the sixteenth of it. (The next lunar month ends) on the sabbath of Jedaiah, on the second of the eleventh (solar) month. Lunar observation takes place [on the third of Seorim, on the nineteenth of it. (The next lunar month ends) on the second (7) of Mijamin, on the second of the twelfth (solar) month.

The Dead Sea Jesus

Lunar observation takes place on the fourth of Abijah, on the eighteenth of it. The fourth (year): (The next lunar month ends) on the fourth of Shecan[iah, on the first of the first (solar) month. Lunar observation takes place on the sixth] (8) [of Jakim, on the seventeenth of the] first (solar) month. (The next lunar month ends) on the Sabbath of Pethahiah, on the thirtieth of the second (solar) month. Lunar observation takes place on the first of Hez[ir, on the seventeenth of it. (The next lunar month ends) on the first of Deliah, on the ninth...]

Fragment 2

Column 1

(1) [Lun]ar observation takes place on the first [of Bilga]h, on the [twenty-] four[th] of it. (The next lunar month ends) on the Sabbath of Hezir, on the seven[th of the sixth (solar) month.] (2) [Lunar observation takes place on the second of Pethahiah,] on the twenty-third of it. (The next lunar month ends) on the first of Jachin, on the fifth of the seventh (,solar) month. Lunar observation takes place on the fourth (3) [of Delaiah, on the twenty-second] of it. (The next lunar month ends) on the third of Joiarib, on the fifth of the eighth (solar) month. Lunar observation takes place on the fifth of Harim, (4) [on the sixteenth of it. (The next lunar month ends) on the fourth of Malachiah, on the fourth of the ninth (solar) month. Lunar observation takes place on the Sabbath of Abijah, on the (5) [twenty-] first [of it. (The next lunar month ends) on the sixth of Je]shua, [on] the third of the tenth (solar) month. Lunar observation takes place on the first of Jakim, on the nineteenth of [it.] (6) [(The next lunar month ends) on the Sabbath of Jeshebeab, on the second of the elev]nth (solar) month. Lunar observation takes place on the thi[rd of Immer, on the nineteenth of it. (7) [(The next lunar month ends) on the second of Happizzez, on the second of the twelfth (solar) month. Lunar observation takes place on the fourth of Jehezkel, on the eighteenth (8) [of it. The first [year: the first month (begins) [in Del]aiah. In Ma[aziah] (9) [is the Passover. In Jedaiah is the Lifting of the Omer. The second month (begins) in] Jedaiah [In Seorim is the Second Passover. The third month (begins) in Hakkoz.]

Column 2

(1) In Je[sh]ua is the Festival of We[eks. The fourth month (begins) in Elia]shib. The fifth month (begins) in Bilgah. The sixth month (begins) in Jehezkel.] The seven[th month (begins) (2) in Maaziah. That day is the Day of Remembrance. In Joiarib is the Day of Atonement. In Jedaiah is the [Festival of] Booths. The eighth month (begins) [in Seorim.] (3) The ninth month (begins) in Jeshua. The tenth month (begins) in Huppah. The eleventh month (begins) in Hezir. The twelfth month (begins) in Gamul. (4) The second (year): the first month (begins) in Jediah. In Seorim is the Passover. In [Mi]jamin is the Lifting of the Omer. The second month (begins) in Mi[jamin. In Abijah] (5) is the [Second] Passover. [The third month (begins) in Eliashib,] and in Hu[ppah] is the Feast of Weeks. [The fourth month (begins) in Bil]gah. The fifth month (begins) in Pethahiah. (6) [The sixth

month (begins) in Maaziah. The seventh month (begins) in Seorim. That day is the Daly of Remembrance. In Malachiah [is the Day of Atonement. In Mijamin (7) [is the Festival of Booths. The eighth month (begins) in Abijah. The ninth month (begins) in] Huppah. The tenth month (begins) in Hezir. The eleventh] month (begins) in Jachin. (8) The [twe]lfth [mon]th (begins) [in Jedaiah The] third (year): The first month (begins) in [Mijam]in. In Abijah is (9) the Passover. In Shecan[ia]h is the Lifting of the Omer. The seco[nd] month (begins) in Shecaniah. In Jakim is the [Se]cond Passover. The third month (begins) in Bilgah. In [He]zir is

Column 3

(1) [the Festival] of Week[s. The [four]th month (begins) in [Pethahiah. The fifth month (begins) in Delaiah. The sixth month (begins) in Se]orim. [The seventh month (begins) in Abijah. That day is the D]ay of [Remembrance. In Jeshua] is the D[ay of Atonement.] (2) [In Shecani]ah is the Festival of [B]ooths. [The] eig[hth month (begins) in Jakim. The ninth month (begins) in Hezir. The tenth month (begins) in] Jachin. The ele[venth month (begins) in] Joiarib [The twelfth] (3) month (begins) in [Mijlamin. [The fourth (year): the first month (begins) in Shecaniah. In Jakim is the Passover. In Jeshebeab is the Waving of the Omer. The second] (4) [month] (begins) in Jeshebeab. In [Immer is the Second Passover. The] third month (begins) [in Pethah]iah. [In Jachin is the Festival of Weeks. The fourth month (begins) in Delaiah] (5) [The fifth month (begins)] in Harim. [The sixth month (begins) in Abijah. The seventh month (begins)] in Jaki[m. That is the Daly of [Remembrance]. In H[uppah is the Day of Atonement. In Jeshebeab is the Festival] (6) [of Booths.] The eighth month (begins) in Immer. [The ninth month (begins) in Jachin. The tenth month (begins) in joiari]b. The elev[en]th [month (begins)] in Ma[lachiah. The twelfth month (begins)] (7) [in Shecaniah.] The fifth (year): the fi[rst month (begins) in Bilgah. In Immer] is the Passover. In Happ[izzez is] the Waving of the Omer. The sec[ond month (begins) in Happizz]ez. In Jehezkel is (8) [the Second Passover. The third month (begins) in [Delaiah. In Joiarib is the Festival of Weeks. The fou[rth month (begins)] in Harim. The fifth month (begins) in Hakk[oz. The sixth month (begins) in Jakim. The seventh month (begins) (9) in Immer. That day is the Day of Rememb[rance. In Hezir is the Day of Atonement. In] Happizzez is the Festival of Booths. The eighth month (begins) in [Je]hezkel. The ninth month (begins) in joiar[i]b. The tenth...

Column 4

(4) [The sixth month (begins) in Immer. The seventh month (begins) in Jehezkel That day is the Day of Remembrance. In Jachin is the Day of Atonement. In Gamul is] the Festival of (5) [Booths. The eighth month (begins) in Maaziah. The ninth month (begins) in Malachiah. The tenth month (begins) in Jeshua The eleven[th mo]nth (begins) in Huppah. (6) [The twelfth month (begins) in Happizzez.]

Priestly Courses 11 (4Q320)

Relying largely on ciphers, Fragment 1 of this manuscript preserves three columns. These portions provide the first three years of the correspondence between the lunisolar calendar and the solar calendar. One can observe that the lunisolar calendar loses ten days per year vis-a-vis the solar calendar, so that over the period covered by these columns it falls behind a full 30 days.

Text

Fragment 1

Column 1

(1)... to show it forth from the East (2)... in the midst of Heaven, in the foundation of (3)... from evening until morning. On the fourth, on the sabbath, (4) the sons of Gamul (shall serve), in the first month, in the first (5) year. (6) [The fifth day of the course of Jedai]ah = the twenty-ninth day (of the lunar month) = the thirtieth day (of the solar month) in it (i.e., in the first month of the solar year). (7) [The Sabbath of Hak]koz (i.e., when the course rotates in; it does not serve until the following sabbath) = the thirtieth = the thirtieth in the second month. (8) [The second of Elia]shib = the twenty-ninth = the twenty-ninth in the third month. (9) [The third of Bilg]ah = the thirtieth = the twenty-ninth in the fourth month (Note: this is a scribal error; the correct equivalence is the twenty-eighth). (10) [The third of Petha]hiah = the twenty-ninth = the twenty-seventh in the fifth month. (11) [The sixth of Deliah] = the thirtieth = the twenty-seventh in the sixth month. (12) [The Sabbath of Seori]m = the twenty-ninth = the twenty-fifth in the seventh month. (13) [The second of Abijah = the thirtieth] = the twenty-fifth in the eighth month. (14) [The third of Jakim = the twenty- [nin]th = the twenty-fourth in the ninth month.

Column 2

(1) The fifth of Immer = the thirtieth = the twenty-third in the tenth month. (2) The sixth of Jehezkel =the twentyninth = the twenty-second in the eleven[th] month. (3) The first in Jehoiarib = the thirtieth = the twenty-second in the twelfth month. (4) The second year: (5) The second of Malakiah (sic!) =the twenty-ninth = the twentieth of the first month. (6) The fourth of Jeshua = the thirtieth = the twentieth of the second month. (7) The fifth of Huppah = the twenty-ninth = the nineteenth of [the sixth month.] (8) The Sabbath of Happizzez = the thirtieth = the eighteenth of the fou[rth month.] (9) The first of Gamul = the twenty-[ninth = the seventeenth of the fifth month.] (10) The third of Jedaiah = the thirtieth = [the seventeenth of the sixth month.] (11) The fourth of Mijamin = the twenty-[nin]th [= the fifteenth of the seventh month.] (12) The sixth of Shecaniah = the thin[ieth = the fifteenth of the eighth month.] (13) The Sabbath of Bil[gah = the twenty-ninth = the fourteenth of the ninth month.] (14) [The second of Pethahiah = the thirtieth = the thirteenth of the tenth month.]

Column 3

(1) [The third of Delaiah = the twenty-ninth = the twelfth of the eleventh month. (2) The fifth of Harim = the thirtieth = the twelfth of the twelfth month. (3) The third year: (4) The fifth of Hakkoz = the twenty-ninth = the twentieth of the first month. (5) The first of Jakim = the thirtieth = the tenth of the second month. (6) The second of Immer = the twenty-ninth = the ninth of the third month. (7) The fourth of Jehezkel = the thirtieth = the eighth of the fourth month. (8) The fifth of Maaziah = the twenty-ninth = the seventh of the fifth month. (9) The Sabbath of Malakiah = the thirtieth = the seventh of the sixth month.] (10) The first of Je[shua = the twenty-ninth = the fifth of the seventh month.] (11) The third of Huppah = the thirtieth = the fifth of the eighth month. (12) The fourth of Hezir = the twenty-ninth = the fourth of the ninth month. (13) The sixth of Jachin = the thirtieth = the third of the tenth month. (14) The Sabbath of Jedaiah = the twenty-ninth = the second of the eleventh month. (15) [The second of] Mijamin = the thirtieth = day [two] of the twelfth month.

Priestly Courses III - Aemilius Kills (Manuscripts A-E - 4Q323-324ª-B)

Though extremely fragmentary, this series of manuscripts provides another record of the proper rotation of priestly courses in the sexennial cycle. This is fairly straightforward and similar to the two preceding ones, though in this one no attempt is made at lunisolar harmonization. It is worth mentioning, too, that its Hebrew is closer to that of early Rabbinic literature than many texts at Qumran.

Text

Manuscript A

Fragment 1

(1)... on the tenth [of the sixth month (i.e., of the second year of the priestly rotation)...] (2) [on the fourteenth of it, the arriva]l of (the priestly course of) Jedaiah; on the sixtee[nth of it... on the twenty-first] (3) [of it the arrival of (the priestly course of) Harim; on the twenty]-seventh of the [sixth] month... (4) he returned... (5) [Gen]tiles and also... (6)... [b]fitter of spirit... (7) prisoners...

Fragment 2

(1) [...to] give him honor among the Arab[s...] (2) [on the fou]rth [day] of this course's service... (3) which is the twentieth of the... month... (4) foundation, Shelamzion came... (5) to greet... (6) Hyrcanos rebelled [against Aristobulus?...] (7) to greet...

The Dead Sea Jesus

Fragment 3

(2) [...the leader of the Gen]ntiles murdered... (3) [on the fi]fth [day] of (the service of the priestly course of) Jedaiah...

Fragment 4

(1)... according to the wi[ll of...]

Manuscript B

Fragment 1

(1)... on the nin[th of the eighth month (i.e., of the second year of the priestly rotation), the arrival of (the priestly course of) Shecaniah...] (2) On day... of (the service of the priestly course of Shecaniah, [...On the sixteenth of it (i.e., the eighth month), the arrival of (the priestly course of) Eliashib]; (3) [on the twenty-thi]rd of it, the arrival of (the priestly course of) Jakim; on the second (day of the service of the priestly course of) Jakim ,... ; and on the fo[urth] day of (the service of the priestly course of) Jakim...] (4) the second day of the ni[nth] month...

Fragment 2

(1) [...the four]th [day] of (the service of the priestly course of) Hez[i]r, [this day is the fi]rst (day) of the te[nth month (i.e., of the second year of the priestly rotation);] (2) [on the fourth day of it (i.e., the tenth month), the arrival of (the priestly course of) Happi]zzez; on the eleventh of it, [the arrival of (the priestly course of) Pethahiah (3) [on the eighteenth of it, the arriv]al of (the priestly course of) Jehezkel; on the twee[ty-fifth of it, the arrival] (4) [of (the priestly course of) Jachin... Jach]in, the se[r]vice... (5) [on the second of it (i.e., the eleventh month)], the arrival of [(the priestly course of) Gamul...]

Fragment 3

(1)... which is... (5) me[n...] (6) and against Ar[istobulus...] (7) [and] they [sa]id... (8) seventy... (9) which is...

Manuscript C

Fragment 1

(1) [on the twenty-third of it (i.e., the fifth month of the fifth year of the priestly rotation)], the arrival of [(the priestly course of) Eliashib; on the thirtieth of it, the arrival of (the priestly course of) Jakim;] (2) after the sabbath, while Jakim is serving, this is the fir]st of the six[th month; on the seventh of it, the arrival of (the priestly course of) Huppah;] (3) [on the four]teenth of it, [the arrival of (the priestly

231

course of) Jeshebeab;]... on the twen[ty-first] (4) [of it, the arrival of (the priestly course of) Bilg]ah; on the twenty-[eighth of i]t, the arrival of (the priestly course of) Imm[er; day] (5) [four (of the service of the priestly course of) Immer is the fi]rst day of the seventh month; on the four[t]h of it, the arrival of (the priestly course of) He[zir;] (6) [the sixth day of] (the service of the priestly course of) Hezir, which is the tenth day of the seventh month, this is [the Day of Atonement;] (7)... for the Covenant; on the eleventh day of the seventh month, the arrival of (the priestly course of) [Happizzez...]

Manuscript D

Fragment 1

Column 2

(5) day... [on the twenty-first] (6) [of i]t (i.e., of the ninth month of the fifth year of the priestly rotation), the arrival of (the priestly course of) S[eor]im; on the twenty-eighth of it, the arrival of (the priestly course of) Malchi[jah;] (7) the fourth day of (the service of the priestly course of) Malchijah is the first day of the tenth month. (8) On the F[ourt]h day of the te[n]th month, the arrival of (the priestly course of) Mijamin; on the eleventh of it, the arrival of (the priestly course of) Hakkoz;]

Fragment 2

(1) [...on the] twenty-(2) [first of it (i.e., of the sixth month of the sixth year of the priestly rotation), the arrival of (the priestly course of) Pethahiah; on the twenty-eigh]th (3) [of it, the arrival of (the priestly course of) Jehezkel; on the first (or, the second; or, the third) day of (the service of the priestly course of) J]ehezkel, which is (4) [the twenty-ninth (or, the thirtieth; or, the thirty-first) day of the sixth month, the Day] of Aemilius' Massacre (literally, 'Aemilius killed' as below); (5) [the fourth day of (the service of the priestly course of) Jehezkel is the first day of] the seventh [mon]th; (6) [on the fourth of it (i.e., of the seventh month), the arrival of the (the priestly course of) Jachin; on the eleventh of it, the arr]iv[al of] (the priestly course of) Gamin; (7) [...the fourth day of (the service of the priestly course of) Gamul, whi]ch is (8) [the fifteenth day of the seventh month, is the Festival of Booths; on that day,] Aemilius killed...

Fragment 3

(2) [on the twenty-eighth of it (i.e., of the ninth month of the sixth year of the priestly rotation), the arrival of (the priestly course of) Je]shua; the four[th] day of [(the service of the priestly course of) Jeshua is the first day of the] (3) [tenth month... wh]ich is the ten[th...]

Fragment 4

(1)... a Jewish man...

Manuscript E

Fragment 1

<u>Column 1</u>

(1)... the hi[g]h priest... (2) Johanan to bring the...

<u>Column 2</u>

(2) from... (5) a man... (7) Shelamzion...

Priestly Courses IV (4Q325)

This manuscript consists of three fragments, the two largest of which are transliterated here. These fragments belong to the first year rotation in the sexennial priestly cycle. The text records which priestly course is responsible for each Sabbath and festival in the period covered.

Text

Fragment 1

(1) [on Tues]day; on the eighteenth the Sabbath fa[lls to Jehoiarib...] (2) [on Tuesday] in the evening. On the twenty-fifth the Sabbath falls to Jedaiah; also during that course's duties falls (3) [the Festival of] the Barley on the twenty-sixth. After the Sabbath, the beginning of the sec[ond] month falls to (4) Jedaiah, [on Fri]day. On the second is the Sabbath of Harim. On the ninth is the Sabbath of (5) [Seorim.] On the sixteenth is the Sabbath of Malchijah. On the twenty-third i[s] (6) [the Sabbath] of [Mi]jamin. On the thirtieth is the Sabbath of Hakkoz. The beginning of (7) the third month, after the Sabbath...

Fragment 2

(1) [the fifth (month) falls to Bilgah. On the second is the Sabbath of I]mmer. On the th[ir]d, (2) [after the Sabbath, is the Festival of New Wine. On] the ninth is the Sabbath of Hezir. (3) [On the sixteenth is the Sabbath of Happizzez. On the twen]ty-third is the Sabbath of (4) [Pethahiah. On the thirtieth is the Sabbath of Jehezkel. The firs]t of the sixth month is (5) [after the Sabbath. On the seventh is the Sabbath of Jachin. On the fo]urteenth (6) [is the Sabbath of Gamul. On the

twentyfirst is the Sabbath of Delaiah. On the twenty-]second (7) [is the Festival of Oil. On the twenty-third is the Festival of W]ood [Offering]...

Heavenly Concordances (OTOT - 4Q319A)

As we have seen, the Qumran calendrical texts are based upon an understanding of the Creation narrative of Genesis. No portion is more significant for these texts than Gen. 1:14, which the authors might have understood as 'Let there be lights in the expanse of heaven to separate the days from the night, and let them be for "signs" (otot), and for festivals and for days and for years.'

Text

Column 1

(10)... its light on the fourth (day); on the sabba[th... (11) the] creation, on the fourth (day) of (the rotation of the priestly course of) Ga[mul, the sign of Shecaniah; in the fourth (year) the sign of Gamul; in the sabbatical year, the sig]n (12) [of Shecaniah; in the thi]rd (year) the sign of [Ga]mul; in the sixth (year), the sign of [Shecaniah; in the second, the sign of Ga]mul; (13) [in the fifth, the sig]n of Shecaniah; after the sabbatical year, the sign of Gam[ul; in the fourth, the sign of Shec]an[ia]h; (14) [in the sabbatical year, the sig]n of Gamul; in the third, the sign of Shecaniah; [in the sixth, the sign of Gam]ul; 0 5) [in the second, the sig]n of She[caniah]; in the fifth, the sign of Gamul; after the sabbatical ye]ar, the sign of (16) [Shecaniah; in the fou]rth, the sign of Gamul; in the sabbatical year, the sign of the con[elusion of the second jubilee. The signs of the [second] jubilee: (17) seventeen signs, of which [three] signs fall in a sabbatical year... the creation (18) [...the sig]n of Sheca[nia]h; in the third year the sign of Gamu[l; in the sixth, the sig]n of Shecaniah; (19) [in the second, the sign of Ga]mul; in the fifth, the sign of Shecaniah; after the sabb[atical year, the sign of Ga]mul;

Column 2

(1) [in the fourth, the sign of Shecaniah; in the sabbatical year, the sign of Gamul; in the third, the sign of Shecaniah;] (2) [in the sixth, the sign of Gamul; in the] sec[on]d, the sig[n of Shecaniah; in the fifth, the sign of Gamul;] (3) after the sabbatical ye]ar, the sign of Shecaniah; in the fou[rth, the sign of Gamul; in the sabbatical year, the sign] (4) of Shecaniah; in the thi]rd, the sign of Gamul; in the six[th the sign] of Shecani[ah; in the second, the sign of the conclusion] (5) of the thi[r]d jubilee. The signs of [the third] jubilee: six[teen signs,] of which (6) two signs fall in a sabbatical year. (A jubilee of) Shecaniah: [in the second year the sig]n of [Gam]ul; in the fifth, the sign of Shecaniah; (7) after the sabbatical year, the sig[n of Gamin; in the fourth, the sig]n of Shecaniah; in the sabbatical year, the sign (8) of Gamul; in the third, the sign [of Shecaniah; in the sixth, the sign of

Ga]mul; in the second, the sign (9) of Shecaniah; in the fifth, the sign [of Gamin; after] the sabbatical year, the sign of Shecaniah; (10) in the fourth, the sign of Gamul; [in the sabbatical year, the sign of] Shecaniah; in the third, the sign of Gamul; (11) in the sixth, the sign of Shecan[iah; in the second, the sign of] Gamul; in the fifth, the sign of Shecaniah; (12) after the sabbatical year, the si[gn of the conclusion of the fourth jubilee. The signs] of the fourth [jubil]ee: seventeen signs, (13) of whi[ch] two signs fall in a sabbatical year. (A jubilee of) Sh[ecaniah:] in the fourth year the sign of Shecaniah; (14) [in the sabbatical] year the sign of Gamul; in the [third, the sign of Shecaniah; in the sixth, the sign of Gamul;] (15) in the seco[n]d, the sign of Shecaniah; in the fifth, the sign of Gamul; after the sabbatical year, the sign of Shecaniah;] (16) in the fourth, the sign of [Ga]mul; in the sabbatical [year, the sign of Shecaniah; in the third, the sign of Gamul;] (17) in the six[th, the si]gn of Sh[ecaniah; in the second, the sign of Gamul; in the fifth, the sign of Shecaniah;] (18) [after the] sabbatical year, the sign of Ga[mul; in the fourth, the sign of Shecaniah; in the sabbatical year, the sign of the conclusion] (19) [of the fifth [jubilee, falling] during (the priestly course of) Jeshebeab. [The signs of the fifth jubilee: sixteen signs, of which]

Column 3

(1) [three signs fall in a sabbatical year. (A jubilee of) Gamul: in the third year, the sign of Shecaniah; in the sixth, the sign] (2) [of Gamul; in the] second, the sign of Shecaniah; in the [fifth, the sign of Gamul; after the sabbatical] year, (3) [the sig]n of Shecaniah; in the fourth, the sign of Gam[ul; in the sabbatical year, the sign of Shecaniah;] in the third, (4) the sign of Gamul; in the sixth, the sign of Shecaniah; [in the second, the sign of] Gamul; (5) in the fi[ft]h, the sign of Shecaniah; after [the sabbatical year,] the sign of (6) Gamul; in the [fo]urth, the sign of Shecaniah; in the sabbatical [year, the sign of Gamul; in the] third, (7) the sign of [Shecaniah; in the sixth, the sign of the conclusion of the [sixth] jubi[lee. The signs (8) of the sixth] jubilee: si[xteen signs,] of which two signs fall in a [sabbatical year...] (9)... (10) And regarding the jubilee of Gamul: in the second year the sign of Shecaniah; in the fifth the sign of Gamul; after] (11) the sabbatical [year the sign of Shecaniah; in the fourth, the sign of Gamu[l; in the] sabbatical ye[ar] (12) [the sign of Shecaniah; in the third, the sign of] Gamul; in the sixth, the sig[n of Shecaniah;] (13) [in the] second, the sig[n of Gamul;] in the fifth, the sign of Shecaniah; [after] (14) the sabbatical yea[r, the sign of Ga]mul; in the fourth, the sign of Shecaniah; in the sabbatical 1 [year, the sign] (15) of Gamul; in the th[ir]d, the sign of Shecaniah; in the sixth, the sign of [Gamul;] (16) in the se[cond, the sign of Shecaniah;] in the fifth, the sign of the conclusion of [the] seventh jubil[ee.] (17) [The signs of the] seventh [jubilee:] sixteen signs, of which (18) [two signs] fall in a sabbatical year... sign of the j[u]bilees, the [y]ear of (the) jubilees according to the day[s of...] (19) in (the priestly course of) Mijamin, the third...

Calendrical Document (Mishmarot)

4Q327 is part of the scrolls known as the calendars. In the calendars, the festivals of the year and the rituals are determined, using priestly rosters. The manuscripts were found in very bad condition. They were also found with several other fragments making it difficult to determine what the remains actually were. Because 4Q327 is in the same handwriting as one manuscript of A Sectarian Manifesto, it is sometimes argued that it should be considered part of that document. Abegg argues against this on the basis of the structure of the latter document. 4Q394 was found in Cave 4 manuscripts. 4Q394 is part of the Halakhic Letter. The Halakhic letter is very important, for it outlines the rules and rituals found in a particular interpretation of the Old Testament. The rest of the works were lost, leaving it unfeasible to determine the true meaning of the Halakhic Letter. Some believe that it was composed as a way to contrast the Qumran group from the rest of Judaism. Each line of the composite text is numbered consequently, for easier reference to the fragments, which have been preserved.

A significant feature of the community was its calendar, which was based on a solar system of 364 days, unlike the common Jewish lunar calendar, which consisted of 354 days. The calendar played a weighty role in the schism of the community from the rest of Judaism, as the festivals and fast days of the group were ordinary work days for the mainstream community and vice versa.

According to the calendar, the new year always began on a Wednesday, the day on which God created the heavenly bodies. The year consisted of fifty-two weeks, divided into four seasons of thirteen weeks each, and the festivals consistently fell on the same days of the week. It appears that these rosters were intended to provide the members of the "New Covenant" with a time-table for abstaining from important activities on the days before the dark phases of the moon's waning and eclipse (duqah). Copied ca. 50-25 B.C.E.

<u>Mishmarot</u>

1. [on the first {day} in {the week of} Jedaiah {which falls} on the tw]elfth in it {the seventh month}. On the second {day} in {the week of} Abiah {which falls} on the twenty- f[ifth in the eighth {month}; and duqah {is} on the third] {day}
2. [in {the week of} Miyamin {which falls} on the twelfth] in it {the eighth month}. On the third {day} in {the week of} Jaqim {which falls} on the twen[ty-fourth in the ninth {month}; and duqah {is} on the fourth] {day}
3. [in {the week of} Shekania {which falls} on the eleven]th in it {the ninth month}. On the fifth {day} in {the week of} Immer {which falls} on the twe[n]ty-third in the te[nth {month}; and duqah {is} on the sixth {day} in {the week of} Je]shbeab {which falls}

4. [on the tenth in] it {the tenth month}. On the [si]xth {day} in {the week of} Jehezkel {which falls} on the twenty-second in the eleventh month [and duqah {is on the} Sabbath in] {the week of} Petahah {which falls}
5. [on the ninth in it {the eleventh month}]. On the first {day} in {the week of} Joiarib {which falls} on the t[w]enty-second in the twelfth month; and [duqah {is} on the seco]nd {day} in {the week of} Delaiah {which falls}
6. [on the ninth in it {the twelfth month}. vacat The] se[cond] {year}: The first {month}. On the sec[on]d {day} in {the week of} Malakiah {which falls} on the tw[entieth in it {the first month}; and] duqah {is}
7. [on the third {day} in {the week of} Harim {which falls} on the seventh] in it {the first month}. On the fou[r]th {day} in {the week of} Jeshua {which falls} [on] the twentieth in the second {month}; and [duqah {is} on the fifth {day} in {the week of]} Haqqos {which falls} on the seventh
8. [in it {the second month}. On the fifth {day} in {the week of} Huppah {which falls} on the nine]teenth in the third {month}; and duqa[h] {is} on the six[th {day} in {the week of} Happisses {which falls}

Fernando Klein

Divination, Magic and Miscellaneous

Brontologion (4Q318)

The present work in Aramaic is perhaps the most intriguing divination text found at Qumran, for it is simultaneously a brontologion, a selenedromion and, apparently, a thema mundi. Each of these terms requires some explanation. A brontologion is a text that attempts to predict the future based upon where within the heavens one hears the sound of thunder (the Greek brontos means 'thunder', hence the name).

Text

Fragment 1

(5) [and on the 7th Sagittarius. On the eighth and the ninth Capricorn. On the tenth and the eleventh Aquarius. On the twelfth and the] thirteenth and the [four]teenth (6) [Pisces. On the fifteenth and the sixteenth Aries. On the seventeenth and the eighteenth Taurus. On the nineteenth, the twentieth and the twenty-[first (7) Gemini. On the twenty-second and the twenty-third Cancer. On the twenty-fourth and the twenty-fifth Leo. On the twenty-sixth], the twenty-seventh and the twentyeighth (8) [Virgo. On the twenty-ninth, thirtieth and thirty-first Libra.] (9) [Tishri On the first and second Scorpio. On the third and fourth Sagittarius. On the fifth, sixth and seventh Capricorn. On the eighth...

Fragment 2

Column 1

(1) and on the thirteenth and the [fo]urteenth Cancer. On the fi[ft]eenth and the sixteenth L[e]o. On the seventeenth [and] the eighteenth (2) Vi[r]go. On the [ni]nteenth, the twentieth and the twenty-first Libra. On the twenty-[second and the twenty-third Scorpio. On the twenty-fourth (3) and the twenty-fifth Sagitt[arius]. On the twenty six[th], the twenty-seve[nth] and the twentyeig[hth Capricorn. On the twenty-n[inth] (4) and the thirtieth Aquari[us]. Shevat On the first and the second [Pisc]es. On [the third and the] fourth (5) [Aries. On] the fifth, [the sixth and the seventh Taurus. On the eig[hth and ninth Gemini.] On the tenth (6) [and eleventh] Cancer. On the twelfth, thirteenth and fourteenth Leo. [On the fifteenth and sixteenth [Virgo]. (7) On the seventeenth and eighteenth Libra. On the nineteenth, [twentieth and twenty-first Scorpio. On the twenty-second (8) [and] twenty-third [Sagitta]rius. On the twentyfourth and twenty-fifth Capricorn. On [the twentysixth], twenty-seventh and twenty-e[i]ghth, (9) Aquarius. On the twenty-ninth and thirtieth Pisces.

Column 2

(1) Adar On the first and the second Aries. On the third and the fourth Taurus. On the fif[th, sixth and seventh Gemini]. (2) On the eighth (and) ninth Canc[er. On the tenth and eleventh L]eo. On the twelfth, thir[teenth and fourteenth] (3) Vi[r]go. On

the fifteenth and six[teenth Libra. On the seventeenth (and) eight[eenth Scorpio]. (4) On the [nin]eteenth, twentieth and twentyfirst Sagitt[arius. On the twenty-second] and twenty-thi[rd Capricorn. On the twenty-[fourth and twenty-fifth] (5) Aquarius. On the twenty-sixth, twenty-[seventh and twenty-eighth Pis[ces]. On the twenty-ni[nth, thirtieth and thirty-first] (6) Aries. [If] it thunders [on a day when the moon is in Taurus], (it signifies) [vain] changes in the wo[rld (?)...] (7) [and] suffering for the cities, and destru[ction in] the royal [co]urt and in the city of dest[ruction (?)...] (8) there will be, and among the Arabs... famine. Nations will plunder one ano[ther...] (9) If it thunders on a day when the moon is in Gemini, (it signifies) fear and distress caused by foreigners and by [...]

Physiognomic Horoscopes (4QCryptic-4Q186, 4QPhysiogn=4Q561)

This text belongs to a widespread type of divination especially well known from Graeco-Roman examples. Writers infer a person's character from movements, gestures of the body, colour, facial expressions, the growth of the hair, the smoothness of the skin, the voice, idiosyncrasies of the flesh, the parts of the body, and the body as a whole. Ancient medicine in particular valued physiognomic signs.

These two texts may represent a variety of divination, known as physiognomy, in which a person's personality or fortune may be read from their physical appearance. They contain what appear to be a series of short body type descriptions which may be intended as a sort of catalogue of physical types which might be useful to the physiognomist. Another possibility is that these are 'prophetic' descriptions of the body types of important biblical or eschatological personages. J. Starky, for example suggests that a related text, 4Q534, is a description of the eschatological Prince of the Congregation, while Vermes (357) sees it as a description of Noah.

4Q186 is Hebrew written in a cypher of sorts. The text is written backwards (left to right) and a mixed alphabet is used (Aramaic square script, Paleo-Hebrew and even Greek characters). 4Q561 is in Aramaic.

4Q186

Frag.1 Col.1

The man, whose head and forehead are wide and curved, [...]but the rest of his head is not [...]

Frag.1 Col.2

...his stone is granite.

He has fixed eyes. He has long and slender thighs, toes, and feet. He was born during the second phase of the moon. His spirit has six parts in the house of light and three parts in the house of darkness. He shall be born under the haunch of Taurus and he will be poor. His animal sign is bull.

Frag.1 Col.3

...and his head...[and his cheeks are] fat. His eyes are terrifying... His teeth are different lengths. His hands and fingers are thick. Each of his thighs is thick and very hairy. His toes are thick and short. His spirit has eight parts in the house of darkness and one in the house of light.

Frag.2 Col.1

His eyes are neither dark nor light. His beard is light and curly. The tone of his voice is soft and gentle. His teeth are fine and well aligned. He is neither tall nor short, but well built. His fingers are thin and long. His thighs are hairless. The soles of his feet and toes are even and well aligned. His spirit has eight parts in the house of light in the second column and one in the house of darkness. His birth sign is...and his animal sign is...

4Q561

Frag. 1 col. I

$_1$ [His ...]... and they will be mixed and sparse. His eyes (will be) $_2$of a medium shade. His nose (will be) a long $_3$ and attractive. And his teeth (will be) straight. And his beard $_4$will be relatively thin. His limbs will be $_5$ in fit condition and niether underweight nor overweight. $_6$... $_7$... his elbows will be strong ... $_8$ husky. And his thighs of [medium] $_9$bulk. And his feet will be [of medium] $_{10}$length. His foot $_{11}$... $_{12}$... $_{13}$... $_{14}$... his shoulder... [medium]... His spirit $_{15}$... $_{16}$... full bodied hair.

Frag. 1 col. II

$_1$The voice will be ... $_2$stern (?)... $_3$it will not strain. $_4$The hair of his beard (will be) plentiful ... $_5$he will be neither fat n[or thin... $_6$And they will be short... $_7$His nails will be strong... $_8$and his height will be ...

Frag. 2

$_2$[... His beard(?)]will be reddish... $_3$...His eyes] will be clear and circular... $_4$...The hair of his hea[d ...

An Amulet Formula Against Evil Spirits (4Q560)

The discovery of this text among the Dead Sea Scrolls is significant not least because it is the earliest known Jewish example, antedating its closest rivals for that honour by several centuries. It is not clear whether this text was part of a larger scroll used as a kind of 'recipe book' by a scribe or magician, or whether it was wrapped up and placed in a case.

This kind of conjuring or incantation is known in apocryphal literature like the Book of Tobit (Chapters 6, 8, and 11) and the Book of Enoch (Chapter 7), both of which are extant in fragments at Qumran. The Mishnah (San. 7:7), like the Old Testament, specifically condemns it.

Text

Fragment 1

Column 1

(1)... heart... (2) a new mother, the punishment of those giving birth, a command (of) evil... (3) the male poisoning demon and the female poisoning-demon (is forbidden) [to] enter the body... (4) [(I adjure you) by the Name of He who for]gives sins and transgression, O fever and chills and heartburn (5) [...and forbidden to disturb by night in dreams or by day] in sleep, the male PRK-demon and the female PRK-demon, those who breach (?)... Fragment 1 Column 2 (2)... before h[im...] (4) before him and... (5) and I adjure you, O spirit... (6) 1 adjure you, O spirit ... (7) upon the earth, in the clouds...

The Era of Light is Coming (4Q462)

This narrative evidently takes as its starting point the prophecy given by Noah in Gen. 9:25-27:
'Cursed be Canaan; he shall be his brothers' meanest slave... Blessed by the Lord my *God* be Shem... May *God* make space for Japheth, and let him live in the tents of Shem.'
Shem, of course, was understood as the ancestor of the Jews, while his brothers were regarded as the ancestors of neighbouring peoples. Noah's words, which were read as prophecy, probably relate to the Davidic period. They predict the ultimate supremacy of the Jews over their neighbours, an inspiring thought in any time of oppression.

Text

Fragment 1

(1) Ham and Japheth... (2) to Jacob and he said... And he remembered... (3) [to give to Israel... Then they said... (4) They went [dress]ed in fine raiment because... (5) [And He gave the sons of Canaan] to Jacob as slaves. In love... (6) [He] gave it (the land) to the Many for an inheritance. The Lord is the Ruler... (7) His Glory which (comes) from (the) One fills the waters and the [land.]... (8) To Him alone belongs the Sovereignty. With Him was the Light; with them (the Angels of Darkness?) and upon us was [the Darkness.]... (9) [He will end the Er]a of Darkness, and the Era of Light is coming, when they (the Angels of Light?) will reign forever. Therefore, [they] said... (10) to Israel, because He was in our midst. In the spring Jac[ob went down to Egypt...](11) [upo]n them, and they were put to forced labor, but they were preserved. They cried out to the Lord... (12) And behold, they were given unto Egypt a second time, at the end of the Kingdom, but they were preserv[ed... (13) And He gave the Holy City into the hand of the inhabit]ants of Philistia and Egypt to spoil and ruin her. [They tore down] her pillars... (14) [She (Jerusalem) ex]changed (her loyalty), exalting Evil, so she will receive the poll[ution of her sins]...(15) and her defiance, so she was hated. In her beauty, her jewelry and her clothing... (16) that which she did to herself. A son of the pollution of Ev[il...] (17) her hatred as she was before she was rebuilt .... (18) But *God* remembered Jerusalem...

He Loved His Bodily Emissions (A Record of Sectarian Discipline - 4Q477)

This text was evidently meant to be the kind of record the Mebakker or Bishop was supposed to keep, as mentioned in the last column of the Damascus Document above, and columns in the extant text, which no doubt preceded it, concerning disciplinary actions and such like 'in the camps'. The presence of one of these records here among the corpus from Qumran not only proves such records were kept, but provides an astonishingly vivid eye-witness testimony to the life that was led in these desert communities preparing for 'the last days' or the 'time of the end'.

Text

Column 1
(1)... the men of the... (2) their soul, and to reprove (3)... the camps of the Many, concerning (4)... rebellion

Column 2
(1) to... (2) which... the Light, knowing[ly...] (3) the Many... Johanan ben... (4) he was short-tempered... his name... And also the Spirit of radience with... . (5) he...And Hananiah Nitos they reproved because he... (6) turned aside the Spirit of

the Commun[ity from the Way], and also to mix the... (7) they repr[oved] [be]cause he... and also because he was not... (8) Furthermore, he loved his bodily emissions... (9) Hananiah ben Shim[... they reproved because...] (10) Furthermore, he loved...

Afterthought

Fernando Klein

Qumran and the final rebellion against the Roman Empire

a. Introduction

When Publius Aelius Hadrianus, known to us as Hadrian, took the reigns of power in 117 CE, he inaugurated - at least at first - an atmosphere of tolerance. He even talked of allowing the Jews to rebuilt the Temple, a proposal that was met with virulent opposition from the Hellenists.

Why Hadrian changed his attitude to one of outright hostility toward the Jews remains a puzzle, but historian Paul Johnson in his History of the Jews speculates that he fell under the influence of the Roman historian Tacitus, who was then busy disseminating Greek smears against the Jews.

Tacitus and his circle were part of a group of Roman intellectuals who viewed themselves as inheritors of Greek culture. (Some Roman nobles actually considered themselves the literal descendants of the Greeks, though there is no historical basis for this myth.) It was fashionable among this group to take on all the trappings of Greek culture. Hating the Jews as representing the anti-thesis of Hellenism went with the territory. Thus influenced, Hadrian decided to spin around 180 degrees. Instead of letting the Jews rebuild, Hadrian formulated a plan to transform Jerusalem into a pagan city-state on the Greek polis model with a shrine to Jupiter on the site of the Jewish Temple.

Nothing could be worse in Jewish eyes than to take the holiest spot in the Jewish world and to put a temple to a Roman god on it. This was the ultimate affront. As bad as this was, the real cause of the revolt seems to have been Hadrian's attempt to follow in the footsteps of the Seleucid Greek Empire 300 years earlier by trying to destroy Judaism. Specifically he targeted Sabbath observance, circumcision, the laws of family purity and the teaching of Torah. An attack against such fundamental commandments of Judaism was bound to trigger a revolt-which it did.

b. The Final War

Hadrian succeeded Trajan as Emperor. In the year C.E. 130, he visited Judea to see the ruins of Jerusalem for himself. Impressed with the beauty of the local, he decided to rebuild Jerusalem, but as a pagan city named Aelia Capitolina. In it would be built a temple but it would be a temple dedicated to the head of the Roman pantheon, Jupiter. Jews had waited a score of years to see the promise of Rome fulfilled and were now both disappointed and insulted. Hadrian, a champion of pagan culture and Hellenistic aesthetics, also issued a ban against castration throughout the Roman Empire. Unfortunately this law encompassed anything that was perceived as mutilation of the genitals, including circumcision. This law was not specifically directed against Jews but it affected them the most.

Rabbi Akiva, the great religious leader of the generation, had believed Rome's promise to rebuild Jerusalem and the Temple. Now, feeling personally betrayed and humiliated by Rome's treachery, he used the influence of his leadership to help stir up rebellion among the Jews of Judea and Galilee. In C.E. 132, a second revolt broke out against Rome in Israel. Unlike the first one in the year C.E. 70, this war had the full support of the Jews of the Diaspora. More importantly, unlike the war of C.E. 70, this revolt was under the leadership of one man - who received the undivided loyalty of the Jewish people. This man was called Simeon bar Kosevah. Rabbi Akiva nick-named him Bar Cochba. This name literally means "Son of a Star"; it is the common midrashic designation for the messiah, based upon the verse in Numbers 24:17-19:

"I shall see him, but not now: I shall behold him, but not nigh: there shall come a Star out of Jacob, and a Sceptre shall rise out of Israel, and shall smite the corners of Moab, and destroy all the children of Sheth. And Edom shall be a possession, Seir also shall be a possession for his enemies; and Israel shall do valiantly. Out of Jacob shall come he that shall have dominion, and shall destroy him that remained of the city."

And this is the name that history has best remembered him by. In giving Simeon this name, Rabbi Akiva, the leading religious light of his generation, was proclaiming him the promised messiah of Israel. Akiva's colleagues vainly tried to point out the rashness of this proclamation but Rabbi Akiva would not retract it.

Bar Cochba's personality does not emerge clearly out of the Jewish sources, other than that he was a stern leader who demanded scrupulousness from his followers in all matters, religious, military, personal. Eusebius reports that Bar Cochba regarded himself as a savior who had descended from heaven like a star.

The war raged on for three years until in C.E. 135, the Romans finally succeeded in driving Bar Cochba's army out of each of its strongholds and in forcing the Jews to surrender. Before his defeat by the Romans, Bar Cochba had an altar erected at the site of the Temple ruins, and had the priests reinstate some of the sacrifices. But after his defeat, Hadrian had the altar torn down, completed the building of Aelia over the site of Jerusalem, caused all the religious leaders, certainly including Akiva, to suffer the death of martyrs, and forbade the teaching of the Torah under penalty of death. No Jew was allowed to enter Jerusalem.

c. Rabbi Akiva

While Shimon Bar Kochba was the military commander of the revolt, the spiritual leader was Rabbi Akiva. He had such faith in Bar Kochba that he believed him to be the Messiah, which, unfortunately, he was not. It was during the Bar Kochba revolt that the 24,000 of Rabbi Akiva's students died in a plague. The rabbis understood this plague to be a result of the students lack of respect for each other, and, despite their high level of intellectual development, their lack of proper moral comportment was fatal. Devastated by the death of his pupils, and the failure of the Bar Kochba revolt, Rabbi Akiva nevertheless persevered and continued teaching his surviving students.

Living in such turbulent times, however, Rabbi Akiva's life was not to end peacefully. Ignoring the Roman prohibitions against the Jewish people and their practices, he was declared a criminal for teaching Torah wherever he could, and was eventually captured by the Romans. Tortured, he called out joyfully: "All my life I've been waiting to fulfill the concept 'You shall love Hashem, your G-d, with all your heart, with all your soul and with all your resources...'[the first paragraph of the Shema] and now I finally have the chance." Rabbi Akiva died a martyr's death.

Jewish outrage at his actions led to one of the single greatest revolts of the Roman Era. Simon Bar Kosiba led the uprising, which began in full force in 132 CE.

For many years, historians did not write very much about Simon Bar Kosiba. But then, archeologists discovered some of his letters in Nahal Hever near the Dead Sea. Some of the letters pertain to religious observance, because his army was a totally religious army. But they also contain a tremendous amount of historical facts. We learn that the Jews participating in the revolt were hiding out in caves. (These caves have also been found - full of belongings of Bar Kosiba's people. The belongings - pottery, shoes, etc. - are on display in the Israel Museum, and the caves, though bare, are open to tourists.)

From the letters and other historical data, we learn that in 132 CE, Bar Kosiba organized a large guerilla army and succeeded in actually throwing the Romans out of Jerusalem and Israel and establishing, albeit for a very brief period, an independent Jewish state. The Talmud (Sanhedrin 97b) states that he established an independent kingdom that lasted for two and half years.

Bar Kosiba's success caused many to believe -- among them Rabbi Akiva, one of the wisest and holiest of Israel's rabbis -- that he could be the Messiah. He was nicknamed "Bar Kochba" or "Son of Star," an allusion to a verse in the Book of Numbers (24:17): "there shall come a star out of Jacob." This star is understood to refer to the Messiah.

Bar Kochba did not turn out to be the Messiah, and later the rabbis wrote that his real name was Bar Kosiva meaning "Son of a Lie" -- highlighting the fact that he was a false Messiah.
At the time, however, Bar Kochba - who was a man of tremendous leadership abilities - managed to unite the entire Jewish people around him. Jewish accounts describe him as a man of tremendous physical strength, who could uproot a tree while riding on a horse. This is probably an exaggeration, but he was a very special leader and undoubtedly had messianic potential, which is what Rabbi Akiva recognized in him.

Jewish sources list Bar Kochba's army at 100,000 men, but even if that is an overestimate and he had half that number, it was still a huge force. United, the Jews were a force to be reckoned with. They overran the Romans, threw them out of the land of Israel, declared independence and even minted coins. That is a pretty unique event in the history of the Roman Empire.

d. The Letters from Simon Bar Kochba

Simon ben Kosiba, surnamed Simon bar Kochba ('son of the star') was a Jewish Messiah. Between 132 and 135, he was the leader of the last resistance against the Romans. After the end of the disastrous rebellion, the rabbis called him 'Bar Koziba', which means 'son of the lie'. In the 1960's, several letters written by Bar Kochba were discovered in caves at Wadi Murabba`at and Nahal Hever. They show that Bar Kochba was a rather authoritarian man.

[1]
Shimeon bar Kosiba to Yehonathan and to Masabala.
Let all men from Tekoa and other places who are with you, be sent to me without delay. And if you shall not send them, let it be known to you, that you will be punished.

[2]
Letter of Shimeon bar Kosiba to Yehonathan, son of Be'ayan:
Peace! My order is that whatever Elisha tells you, do to him and help him and those with him. Be well.

[3]
Shimeon to Yehudah bar Menashe in Qiryath 'Arabaya.
I have sent to you two donkeys, and you must send with them two men to Yehonathan, son of Be'ayan and to Masabala, in order that they shall pack and send to the camp, towards you, palm branches and citrons. And you, from your place, send others who will bring you myrtles and willows. See that they are tithed and sent them to the camp. The request is made because the army is big. Be well.

[4]
From Shimeon bar Kosiba to the men of En-gedi.
To Masabala and to Yehonathan bar Bey'ayan, peace!
In comfort you sit, eat and drink from the property of the House of Israel, and care nothing for your brothers.

This last letter seems to be a reproach to the men of En-gedi, because they had failed to take part in a battle.

Bibliography

Badia, Leonard Francis, The Dead Sea People's Sacred Meal and Jesus' Last Supper. Washington: University Press of America, 1979.
Berger, Klaus, The Truth under Lock and Key? Jesus and the Dead Sea Scrolls. Trans. Currie, James S. 1st ed. Louisville, Ky.: Westminster John Knox Press, 1995.
Betz, Otto, and Rainer Riesner, Jesus, Qumran, and the Vatican: Clarifications. Trans. John Bowden. New York: Crossroad, 1994.
Black, Matthew, The Scrolls and Christian Origins: Studies in the Jewish background of the New Testament. Brown Judaic Studies 48. Chico, Calif.: Scholars Press, 1983.
Brooke, George J., Exegesis at Qumran : 4QFlorilegium in its Jewish Context. JSOTSup 29. Sheffield: JSOT Press, 1985.
Broshi, Magen. "The Archeology of Qumran–A Reconsideration." The Dead Sea Scrolls. Forty Years of Research. Ed. Devorah Dimat and Uriel Rappaport. Studies on the Texts of the Desert of Judah 10. Leiden: Brill, 1992. 103-115.
Bruce, F. F., Biblical Exegesis in the Qumran Texts. London: Tyndale Press, 1960.
Calaway, Phillip R., The History of the Qumran Community: An Investigation. JSPSup 3. Sheffield: JSOT Press, 1988.
Cansdale, Lena. Qumran and the Essenes: A Re-Evaluation of the Evidence. Mohr, 1997
Charlesworth, J. "The Origin and Subsequent History of the Authors of the Dead Sea Scrolls: Four Transitional Phases among the Qumran Essenes," RevQ 10 (1980) 213-233.
Charlesworth, James H. (ed.). Jesus and the Dead Sea Scrolls. ABRL. Doubleday, 1992.
Charlesworth, James H. and Walter P. Weaver, ed. The Dead Sea Scrolls and Christian Faith. In Celebration of the Jubilee Year of the Discovery of Qumran Cave 1. Faith and Scholarship Colloquies. Harrisburg, Pa.: Trinity Press International, 1998.
Charlesworth, James H., and Raymond Edward Brown, John and the Dead Sea Scrolls. Christian Origins Library. New York: Crossroad, 1991.
Charlesworth, James H., Qumran Questions. The Biblical Seminar 36. Sheffield: Sheffield Academic Press, 1995.
Collins, John J. "The Origin of the Qumran Community: A Review of the Evidence." To Touch the Text. Ed. M. Horgan and P. Kobelski. New York: Crossroad, 1989. 159-178.
Collins, John J., "Dead Sea Scrolls," ABD 2.85-101.
Collins, John J., "Essenes," ABD 2.619-626.
Cook, Edward M., Solving the Mysteries of the Dead Sea Scrolls. Grand Rapids: Zondervan, 1994.
Cross, Frank M., The Ancient Library of Qumran and Modern Biblical Studies. 3rd ed. Minneapolis: Fortress, 1995.
Davies, Philip R. "The Prehistory of the Qumran Community." The Dead Sea Scrolls. Forty Years of Research. Ed. Devorah Dimat and Uriel Rappaport. Studies on the Texts of the Desert of Judah 10. Leiden: Brill, 1992. 116-125.

Davies, Philip R., Behind the Essenes: History and Ideology in the Dead Sea Scrolls. BJS 94. Atlanta: Scholars Press, 1987.
Dimant, Devorah. "Qumran Sectarian Literature." Jewish Writings of the Second Temple Period. Ed. M. E. Stone. The Literature of the Jewish People in the Period of the Second Temple and the Talmud. Philadelphia: Fortress, 1984. 2.483-550.
Dimant, Devorah. "The Qumran Manuscripts: Contents and Significance." Time to Prepare the Way in the Wilderness. Papers on the Qumran Scrolls by Fellows of the Institute for Advanced Studies of the Hebrew University, Jerusalem, 1989-1990. Ed. Devorah Dimant and Lawrence H. Schiffman. Studies on the Texts of the Desert of Judah 16. Leiden: Brill, 1995. 23-58.
Evans, Craig A., and Stanley E. Porter, The Scrolls and the Scriptures: Qumran Fifty Years After. JSPSup 26. Sheffield: Sheffield Academic Press, 1997.
Evans, Craig A., Noncanonical Writings and New Testament Interpretation. Peabody, Mass.: Hendrickson Publishers, 1992.
Fitzmyer, Joseph A. Responses to 101 Questions on the Dead Sea Scrolls. New York: Paulist, 1992
Fujita, Neil S., A Crack in the Jar: What Ancient Jewish Documents Tell Us About the New Testament. New York: Paulist Press, 1986.
Golb, Norman, "The Dead Sea Scrolls: A New Perspective," American Scholar 58 (1989) 177-207.
Golb, Norman, Who Wrote the Dead Sea Scrolls? The Search for the Secret of Qumran. Toronto: Scribner, 1995.
Horgan, Maurya P., Pesharim: Qumran Interpretations of Biblical Books. CBQMS 8. Washington: Catholic Biblical Association of America, 1979.
Kampen, John, and Moshe J. Bernstein, ed. Reading 4QMMT: New Perspectives on Qumran Law and History. SBL Symposium Series 2. Atlanta, Ga.: Scholars Press, 1996.
LaSor, William Sanford, The Dead Sea Scrolls and the New Testament. Grand Rapids: Eerdmans, 1972.
Lim, Timothy H. Holy Scripture in the Qumran Commentaries and Pauline Letters. Oxford, 1997.
Lim, Timothy, "The Qumran Scrolls: Two Hypotheses," Studies in Religion 21 (1992) 455-466.
Magness, Jodi, "The Chronology of the Settlement at Qumran in the Herodian Period," Dead Sea Discoveries 2 (1995) 58-65.
Martínez, F. García, "A 'Groningen' Hypothesis of Qumran Origins and Early History," RevQ 14 (1990) 536-541.
Martínez, F. García, "Qumran Origins and Early History: A 'Groningen' Hypothesis," Folia Orientalia 25 (1988) 113-136.
Milik, J. T., Ten Years of Discovery on the Wilderness of Judaea. Trans. J. Strugnell. Naperville, Ill: Alec R. Allenson, 1959.
Murphy-O'Connor, J. "The Essenes and their History," RB 81(1974) 215-244.
Murphy-O'Connor, J. "The Essenes in Palestine," BA (Sept. 1977) 100-124.
Murphy-O'Connor, J., "Qumran, Khirbet," ABD 5.590-594.
Schiffman, Lawrence, Reclaiming the Dead Sea Scrolls. New York: Doubleday, 1995.

Schiffman, Lawrence. "Qumran and Rabbinic Halakhah." Jewish Civilization in the Hellenistic-Roman Period. Ed. Sh. Talmon. JSPSup 10. Sheffield: JSOT Press, 1991. 138-146.

Schiffman, Lawrence. "The Sadducean Origins of the Dead Sea Scroll Sect." Understanding the Dead Sea Scrolls. Ed. H. Shanks. New York: Vintage Books, 1992. 35-49.

Shanks, Hershel. "Qumran Origins – Palestine or Babylonia?" Understanding the Dead Sea Scrolls. Ed. H. Shanks. New York: Vintage Books, 1992. 79-86.

Skehan, Patrick W. "Littérature de Qumran." Supplément au Dictionnaire de la Bible. Ed. H. Cazelles and A. Feuillet. Paris: Letouzey & Ané, 1973. 9.805-822.

Stegemann, Hartmut, The Library of Qumran. Grand Rapids/Leiden: Eerdmans/Brill, 1998.

Stegemann, Hartmut. "The Qumran Essenes." The Madrid Qumran Congress. Proceedings of the International Congress on the Dead Sea Scrolls: Madrid 18-21 March, 1991. Ed. J. T. Barrera and L. V. Montaner. STDJ 11,1. Leiden: Brill, 1992. 138?-166.

Stendahl, Krister, The Scrolls and the New Testament. Westport, Conn.: Greenwood Press, 1975.

Thiering, B. E., Jesus & the Riddle of the Dead Sea Scrolls: Unlocking the Secrets of His Life Story. Toronto: Doubleday, 1992.

VanderKam, James C., The Dead Sea Scrolls Today. Grand Rapids: Eerdmans, 1994.

Vaux, Roland de, Archaeology and the Dead Sea Scrolls. London: Oxford University Press, 1973.

Vermes, Geza, and Martin D. Goodman, ed. The Essenes According to Classical Sources. Sheffield: JSOT Press, 1989.

Yadin, Yigael. "The Temple Scroll – The Longest Dead Sea Scroll." Understanding the Dead Sea Scrolls. Ed. H. Shanks. New York: Vintage Books, 1992. 87-112.